Carbonaceous Materials and Future Energy

Carbonaceous Materials and Future Energy
Clean and Renewable Energy Sources

Ramendra Sundar Dey
Taniya Purkait
Navpreet Kamboj
Manisha Das
Institute of Nano Science and Technology
Mohali, India

CRC Press
Taylor & Francis Group
Boca Raton London New York

CRC Press is an imprint of the
Taylor & Francis Group, an **informa** business

CRC Press
Taylor & Francis Group
6000 Broken Sound Parkway NW, Suite 300
Boca Raton, FL 33487-2742

© 2020 by Taylor & Francis Group, LLC
CRC Press is an imprint of Taylor & Francis Group, an Informa business

No claim to original U.S. Government works

Printed on acid-free paper

International Standard Book Number-13: 978-0-8153-4788-0 (Hardback)

This book contains information obtained from authentic and highly regarded sources. Reasonable efforts have been made to publish reliable data and information, but the author and publisher cannot assume responsibility for the validity of all materials or the consequences of their use. The authors and publishers have attempted to trace the copyright holders of all material reproduced in this publication and apologize to copyright holders if permission to publish in this form has not been obtained. If any copyright material has not been acknowledged please write and let us know so we may rectify in any future reprint.

Except as permitted under U.S. Copyright Law, no part of this book may be reprinted, reproduced, transmitted, or utilized in any form by any electronic, mechanical, or other means, now known or hereafter invented, including photocopying, microfilming, and recording, or in any information storage or retrieval system, without written permission from the publishers.

For permission to photocopy or use material electronically from this work, please access www.copyright.com (http://www.copyright.com/) or contact the Copyright Clearance Center, Inc. (CCC), 222 Rosewood Drive, Danvers, MA 01923, 978-750-8400. CCC is a not-for-profit organization that provides licenses and registration for a variety of users. For organizations that have been granted a photocopy license by the CCC, a separate system of payment has been arranged.

Trademark Notice: Product or corporate names may be trademarks or registered trademarks, and are used only for identification and explanation without intent to infringe.

Library of Congress Control Number: 2019952556

Visit the Taylor & Francis Web site at
http://www.taylorandfrancis.com

and the CRC Press Web site at
http://www.crcpress.com

Dedicated to all my family members
(*Ma, baba, Manua and Rishaan*)

Contents

Preface xiii
Acknowledgement xv
Authors xvii
Introduction xix

Chapter 1 Clean and Renewable Energy Sources: Global Comparison 1

    1.1   Introduction 1
    1.2   Non-Renewable Energy Resources 2
            1.2.1   Coal 3
            1.2.2   Oil 3
            1.2.3   Natural Gas 3
    1.3   Renewable Energy Sources 3
            1.3.1   Solar Energy 4
            1.3.2   Wind Energy 4
            1.3.3   Waste and Bioenergy 4
    1.4   Role of Carbon Nanomaterials 5
    1.5   Conclusion 5
    References 6

Chapter 2 Different Allotropes of Carbon, Their Structures and Properties 7

    2.1   Introduction and General Information 7
    2.2   The Element Carbon and Its Allotropes 7
            2.2.1   Graphite: Most Used Carbon Allotropes in Energy 9
            2.2.2   Graphene and Reduced Graphene Oxide 9
            2.2.3   Three-Dimensional Reduced Graphene Oxide 10
            2.2.4   Carbon Nanotubes 11
            2.2.5   Fullerene 12
            2.2.6   Activated Carbon 13
            2.2.7   Carbon Nanofiber 13
            2.2.8   Other Forms of Carbon 14
                      2.2.8.1   Carbon Nano-Onions 14
                      2.2.8.2   Carbon Nanocones 14
    2.3   Summary 16
    References 16

Chapter 3 Synthesis and Characterisation of Carbonaceous Materials 19

    3.1   Introduction 19
    3.2   Synthesis of Graphene 20
            3.2.1   Mechanical Exfoliation 20

		3.2.2	Chemical Vapor Deposition: Synthesis and Transfer	20
			3.2.2.1 Plasma-Enhanced Chemical Vapor Deposition	21
			3.2.2.2 Transfer of Graphene Sheet	21
		3.2.3	Arc Discharge Method	22
		3.2.4	Epitaxial Growth	22
		3.2.5	Chemical Synthesis	22
			3.2.5.1 Synthesis of Graphite Oxide	22
			3.2.5.2 Reduction of Graphite Oxide	23
		3.2.6	Electrochemical Route	23
		3.2.7	Biomass-/Waste-Derived Graphene	23
		3.2.8	Other Synthesis Methods	24
	3.3	Synthesis of Carbon Nanotubes		24
		3.3.1	Arc Discharge	24
		3.3.2	Laser Ablation	25
		3.3.3	Chemical Vapor Deposition Method	25
	3.4	Characterisations		26
		3.4.1	Microscopic Technique	26
		3.4.2	Raman Analysis	27
		3.4.3	X-ray Diffraction	27
		3.4.4	X-ray Photoelectron Spectroscopy	27
	3.5	Conclusion		28
	References			28
Chapter 4	Rechargeable Battery Technology			35
	4.1	Introduction		35
	4.2	Li-Ion Batteries		37
		4.2.1	Cathode Material	39
			4.2.1.1 $LiCoO_2$	39
			4.2.1.2 $LiMn_2O_4$	39
			4.2.1.3 $LiFePO_4$	39
		4.2.2	Anode Material	40
			4.2.2.1 Graphite	40
	4.3	Sodium-Ion Batteries		40
	4.4	Magnesium-Ion Batteries		41
	4.5	Metal-Sulfur Batteries		41
	4.6	Metal-Air Batteries		42
	4.7	Summary and Outlook		43
	References			44
Chapter 5	Supercapacitor			49
	5.1	Introduction		49
	5.2	Charge Storage Mechanism in Supercapacitors		50

		5.2.1	Electrical Double-Layer Capacitor	50
		5.2.2	Pseudocapacitors	51
		5.2.3	Hybrid Supercapacitors	52
	5.3	Device Architecture of Supercapacitor System		52
		5.3.1	Sandwich-Type Supercapacitor	52
		5.3.2	In-Plane Supercapacitor	52
	5.4	Key Parameters for Supercapacitor Performance Evaluation		53
		5.4.1	Specific Capacitance	53
		5.4.2	Energy Density and Power Density	54
		5.4.3	Cycling Stability	55
	5.5	Electrode Materials for Supercapacitors		55
		5.5.1	Carbon-Based Materials	55
			5.5.1.1 Activated Carbons	55
			5.5.1.2 Biomass-Derived Nanocarbon	56
			5.5.1.3 Graphene	57
			5.5.1.4 Other Carbonaceous Structures	58
		5.5.2	Hybrid Materials	58
			5.5.2.1 Nanocarbon-Conducting Polymer Composites	58
			5.5.2.2 Nanocarbon–Metal Oxide Hybrids	59
	5.6	Applications of Supercapacitors		59
	5.7	Future Perspective: Recent Advances and Disadvantages		60
		5.7.1	Advances of Supercapacitors	60
		5.7.2	Disadvantages of Supercapacitors	61
	References			61

Chapter 6 Solar Cell ... 67

	6.1	Introduction	67
	6.2	Different Types of Photovoltaic Devices	67
	6.3	Silicon-Based Solar Cells	68
	6.4	Polymer-Based Solar Cells	69
	6.5	Dye-Sensitised Solar Cells	72
	6.6	Quantum Dots–Based Solar Cells	72
	6.7	Perovskites Solar Cells	73
	6.8	Future Perspectives	74
	References		74

Chapter 7 Fuel Cell ... 77

	7.1	Introduction to Fuel Cell Technology		77
	7.2	Working Principle and Mechanism		78
	7.3	Classification of Fuel Cell		80
		7.3.1	Proton Exchange Membrane Fuel Cell	80
		7.3.2	Alkaline Fuel Cell	81

		7.3.3	Solid Oxide Fuel Cell	81
		7.3.4	Direct Methanol Fuel Cell	81
		7.3.5	Phosphoric Acid Fuel Cell	82
		7.3.6	Molten Carbonate Fuel Cell	82
	7.4	Limitation of Metal-Based Catalyst		82
	7.5	Carbon-Based Catalyst		82
		7.5.1	Carbon-Based Anode Catalyst	82
		7.5.2	Carbon-Based Cathode Catalyst	84
	7.6	Conclusion		85
	References			85

Chapter 8 Electrocatalysis and Photocatalysis ... 89

	8.1	Introduction		89
	8.2	Water Splitting		91
		8.2.1	Electrochemical Water Splitting	93
		8.2.2	Photochemical Water Splitting	101
		8.2.3	Photoelectrochemical Water Splitting	105
	8.3	Carbon Dioxide Reduction		106
		8.3.1	Electrochemical CO_2 Reduction	108
		8.3.2	Photochemical CO_2 Reduction	110
		8.3.3	Photoelectrochemical CO_2 Reduction	113
	8.4	Conclusion		114
	References			114

Chapter 9 Nanogenerator ... 119

	9.1	Introduction			119
		9.1.1	Mechanical Energy Resources		120
		9.1.2	Harvesting Mechanical Energy: Mechanism and Materials		121
	9.2	Classification of Nanogenerators			122
		9.2.1	Piezoelectric Nanogenerators		123
			9.2.1.1	Mechanism	123
			9.2.1.2	Geometrical Configuration	125
			9.2.1.3	Formulae Used	126
			9.2.1.4	Synthesis Methodologies	127
			9.2.1.5	Factors Affecting Performance of Piezoelectric Nanogenerator	127
		9.2.2	Pyroelectric Nanogenerator		128
			9.2.2.1	Mechanism	128
			9.2.2.2	Formulae Used	129
		9.2.3	Triboelectric Nanogenerator		130
			9.2.3.1	Mechanism	130
			9.2.3.2	Formulae Used	131

		9.2.3.3	Boosting Triboelectric Nanogenerator Performance	131
		9.2.3.4	Recent Progress in Advancement in Triboelectric Nanogenerator Applications	132
	9.4	Carbon-Based Nanogenerators		132
	9.5	Conclusions		133
	References			134

Chapter 10 Carbonaceous Materials–Based Hybrid Energy Technologies 141

    10.1 Introduction to Hybrid Technology 141
    10.2 Supercapacitor-Battery Hybrid Devices 143
        10.2.1 Lithium-Ion Capacitor 144
        10.2.2 Sodium-Ion Capacitor 145
    10.3 Supercapacitor-Nanogenerator Hybrid Technology 145
    10.4 Hybrid Device of Solar Cell with Energy Storage System 147
    10.5 Supercapacitor-Biofuel Cell Hybrid Technology 148
    10.6 Hybrid Self-Powered Water Electrolysis Technologies 149
    10.7 Hybrid Technology with Fuel Cell and Energy Storage Devices 151
    10.8 Other Hybrid Technologies 151
    10.9 Summary and Outlook 151
    References 152

Chapter 11 State-of-the-Art Renewable Energy Technology 157

    11.1 Introduction 157
    11.2 Various Renewable Energy Technologies 158
        11.2.1 State-of-the-Art Storage Technologies 158
        11.2.2 Solar Energy Technology 159
        11.2.3 Wind Energy 160
        11.2.4 Water Splitting 161
        11.2.5 Waste to Energy 161
    11.3 Conclusion 162
    References 162

Chapter 12 Future Perspective 167

Index 169

Preface

Worldwide energy depletion is accelerating at an alarming speed. There are three main origins: rapid economic growth, population evolution, and increased dependence on energy-based utilities throughout the world.

The modern lifestyle demands a steady and consistent supply of energy for our daily requisites and comfort. All energy sources have some undesirable characteristics. At present most of the world's energy supply comes from fossil and nuclear sources, which are limited. Still, these sources will continue to be important in providing energy worldwide for the next few generations. Therefore, mankind is increasingly facing issues of resource limitation and environmental pollution. Energy experts have estimated that by 2050, the total supply of energy will need to be doubled.

To date, fossil fuels (coal, oil and gas) are the main source of energy to meet the demand of various industrial and domestic applications. However, the production of energy from such fossil fuels is neither sustainable nor environmentally friendly. In addition, during burning these emit various pollutants including the greenhouse gases (e.g. carbon dioxide), which expedites the process of global warming and negatively impacts the global climate. It is important to note that around 80 wt% of CO_2 emissions in the world are accountable by the energy division. Therefore, to face increasing global demands for energy and to permit for the diminution of fossil fuel supplies in the coming years, alternative 'clean' energy sources are required, which do not depend on fossil fuels and which have bearable environmental impact. This means we have an urgent need for efficient, sustainable and clean sources of energy, as well as new technologies to generate a sustainable chain of energy supply, associated with environmentally friendly and cost-effective energy resources. To achieve this goal, the use of renewable energy can be considered as a prime step. Some of the popular forms of renewable energies are solar energy, hydropower, biomass, biofuel, geothermal, wind power, etc.

Concentrating on the area of energy, nanotechnology has the potential to solve the current complications of energy production and to provide energy opportunities without compromising environmental and human health. Nanotechnologies may lead to smart generation, storage and transport of energy in a clean, sustainable and more efficient way. Precisely, carbon allotropes and derivatives have been widely investigated over the past few years for clean energy generation and storage.

Carbonaceous materials are very different in structure and properties. Various forms of carbonaceous material such as graphene, carbon nanotubes, fullerenes, mesoporous carbon, carbon nanofibers and their composites are extensively studied materials due to their large surface area, excellent mechanical strength, controllable porosity, sufficient active sites and high conductivity. To attain high performance and good mechanical stability, hybrid nanostructure composites of carbonaceous materials in conjunction with electroactive conducting polymers and/or metal oxides have been developed.

This book reviews the roadmap of various carbonaceous materials used in several energy devices and provides a guideline for future perspective. The advantages of various forms of carbonaceous materials used in energy devices and their comparative study are also reviewed. Current and future perspectives of carbonaceous material use in energy applications are presented, which are gathered from many different disciplines. Our hope is that this book on the prospects and potential of various materials in the development of renewable energy sources will significantly enrich the knowledge of researchers searching for viable renewable energy solutions.

Ramendra Sundar Dey
Taniya Purkait
Navpreet Kamboj
Manisha Das
INST, Mohali, India

Acknowledgement

This book will remain incomplete without thanking those people who helped me during this venture.

I am very grateful to my family, my wife Manua and my son Rishaan, for constant moral support, love and faith in me. My words fail to express my gratitude and affection to my parents and my elder brother for their continuous support in my academic pursuits and help to get to this point.

It is my privilege to mention a few friends and colleagues, Dr. De Sarkar and Dr. Hazra from INST Mohali and Director, INST Mohali, for moral support.

I apologize to all those not named, either due to space limitations or imperfections of human memory.

Finally, thanks to the Almighty God for blessings.

Ramendra Sundar Dey
INST, Mohali, India

Authors

Ramendra Sundar Dey, PhD, (Orcid.org/0000-0003-3297-1437) is a scientist at the Institute of Nano Science and Technology, Mohali, India. Prior, he was a Hans C. Ørsted postdoc fellow at Technical University of Denmark (DTU), Denmark. He earned his BSc degree in Chemistry from Burdwan Raj College (Burdwan University) in 2005, MSc degree in Chemistry from the University of Burdwan, India, in 2007 and PhD in Chemistry in 2013 from the Indian Institute of Technology (IIT), Kharagpur, India. Over the past 10 years he has been involved in research in the field of electroanalytical chemistry in nanotechnology. His current research focuses on the architecture and engineering of carbonaceous materials and their applications in advanced energy storage and conversion technology and sensing/biosensing devices.

Dr. Dey has published more than 25 research papers in peer-reviewed international journals of repute and two book chapters and filed one patent. He has been honored with a number of prestigious national and international awards, such as a PhD fellowship from UGC, India (2008–2013), several national awards from India, INSPIRE Faculty Fellowship from DST, India, and the prestigious Hans Christian Ørsted Postdoctoral Fellowship (2013) from DTU, Denmark. He currently serves as a guest editor and editorial advisory board member for a few journals of international repute.

Taniya Purkait is currently earning her PhD in nanoscience and nanotechnology from the Institute of Nano Science and Technology (INST), Mohali, and CNSNT, Panjab University under the guidance of Dr. Ramendra Sundar Dey. She earned her BSc degree from Thakurpukur Vivekananda College (University of Calcutta) in 2012 and MSc degree from the Indian School of Mines (ISM), Dhanbad, in 2014. For the past four years she has been involved in active research in the field of electroanalytical chemistry. Her current research interests include synthesis and fabrication of various carbonaceous materials for various energy storage and conversion applications. She has published seven research papers in reputed peer-reviewed journals.

Navpreet Kamboj is currently earning her PhD in nanoscience and nanotechnology from the Institute of Nano Science and Technology (INST), Mohali, and IISER, Indian Institute of Science Education and Research, Mohali, under the guidance of Dr. Ramendra Sundar Dey. She completed her BSc degree from Govt. P. G. College (Kurukshetra University, Kurukshetra) in 2010 and MSc degree in physics from Kurukshetra University, Kurukshetra, in 2012. She has worked as a teaching assistant in Govt. P. G. College, Hisar, from 2013 to 2015. She earned her MTech degree in Material Science from IIT Delhi in 2017. She has been involved in active research in the broad field of energy devices. Her current research interest includes synthesis and fabrication of various carbonaceous-based hybrid storage and conversion devices in energy application.

Manisha Das is currently pursuing her PhD in nanoscience and nanotechnology from the Institute of Nano Science and Technology (INST), Mohali, India, and the Indian Institute of Science Education and Research, Mohali, India. She earned her BTech degree in Electronics and Communication Engineering in 2013 from the Sam Higginbottom University of Agriculture, Technology and Sciences (SHUATS), Allahabad, India, and MTech in Material Science and Nanotechnology from National Institute of Technology (NIT), Kurukshetra, India, in 2017. She has published three research papers in peer-reviewed journals. Her areas of interest are nanomaterials synthesis and electrocatalysis for energy generation and conversion.

Introduction

I. BACKGROUND AND OVERVIEW OF TOPICS

In science class, we were taught that energy can neither be created nor destroyed, but it can be transformed from one variety to another. Energy and the environment are presently demanding enormous awareness both academically and technologically to build a green society. Continuous increases in the global population and speedy technological developments are major factors responsible for many environmental issues [1]. Fossil fuels are the current existing source of energy for world energy demand. Therefore, limited sources of fossil fuels and the exponential growth of the population are driving factors in the energy crisis. Most energy experts predict that we need an additional 30 trillion watts of extra energy by 2050, which is almost three times more than current energy consumption [2].

It is essential to look for potential sustainable, renewable and green energy sources to substitute for the age-old and limited fossil fuel sources, which is the most important and thought-provoking issue in energy research [3]. It is universally well known that because of air and water pollution, climate change and biodiversity, together with oil depletion due to the use of fossil fuels, researchers are forced to dedicatedly look for alternative solutions. There are various renewable energy technologies including energy storage and conversion devices which promise sustainable solutions [4]. Nanotechnology offers a different solution for finding new cost-effective materials and efficient technology and having less environmental impact in creating sustainable energy storage and conversion systems.

In this circumstance, carbon plays an important role. Carbon is the most abundant material on earth and has a huge number of allotropes in the forms of carbon black, activated carbon, fullerene, carbon nanotubes, carbon nanofiber and graphene. They are regularly utilized for renewable energy applications due to their many pleasing properties such as surface area, electrical and ionic conductivity, mechanical flexibility and cost-effectiveness [5].

II. FUTURE ENERGY: OPPORTUNITY AND CHALLENGES

In the near future, all researchers will be forced to focus on the development of new and renewable energy sources. The massive uses and the limited sources of fossil fuels will lead to an energy crisis in this century. The development of alternative methods for searching for renewable energy sources is alarmingly necessary to survive as well as to improve and maintain the living standard for mankind. Future energy needs to be generated from solar, hydro, wind geothermal, mechanical nuclear power and waste energy harvesting technologies. Recently it has been observed that human, magma, tidal, hydrogen and even algae power can play crucial roles in the generation of renewable energy for future generations. Specifically, the world is populated with seven billion people, and power can be generated from walking humans, which is power that is otherwise being wasted. Algae grow almost everywhere, and these tiny plants contain energy that may be possible to convert to various biofuels.

There are many reports that debate the feasibility of achieving 100% renewable energy in the near future. And there is a lack of any historical evidence of the technical feasibility of achieving 100% renewable electricity, both globally and regionally [6,7]. In a broader context, resource constraints and lack of technological maturity, socio-economic viability and laws of governing have been setbacks to achieving it globally. Therefore, only a few of the most popular versions of renewable energies that are really making substantial progress are briefly discussed here, highlighting recent opportunities and developments.

Among all renewable energy resources, solar energy is the most promising and highest used alternative energy. This is due to the fact that the most abundant power source on the planet is sunlight. The total 174 petawatts (PW) of incoming solar radiation are being received by the earth in the upper atmosphere. Among them, nearly 30% is reflected back to space while the rest is absorbed by the earth. Considering the world's energy demand, solar cells must be inexpensive and highly efficient in converting solar energy to electricity. The present drawback of the solar cell comes from its conversion efficiency of sunlight, which is due to the unavailability of sunlight throughout the day at a consistent intensity as well as weather conditions and geographical positions [8]. In addition, the architecture and efficiency of the material as well as the device could play crucial roles in the conversion efficiency of the sunlight. The highest efficiency of 'one sun' solar cell achieved to date is 25.6% with crystalline Si, to the best of our knowledge [9]. However, multijunction solar cells have reached record-breaking efficiency of more than 40% in recent years [10]. The crystal dislocations in the metamorphic multijunction solar cell and the shape of the solar spectrum are favourable to achieving high efficiency in the multijunction solar cell designed for terrestrial use.

Continuous research and development in advanced multijunction solar cell materials and device architectures, conducted globally, have influenced researchers to aim for efficiencies up to 50% with concentrator-based solar cells. A powerful demand to bring down the cost of solar cells together with targets for higher efficiency will make it possible to use solar-cell technology on a larger scale than thought before.

Wind energy is another renewable energy technology that is expected to contribute an important role in future energy resources due to its capability to generate power through wind with technological maturity and relative cost effectiveness [11]. As the wind is free in earth's atmosphere and it is renewable, wind energy becomes one of the best alternative non-conventional energy resources due to the following reasons [12,13]: (1) Wind is a low-density fluid, and its kinetic energy changes to other types of energy. (2) The speed of the wind and its availability is a sufficient and non-ending process, although the direction and amplitude are unpredictable. (3) Conversion of wind is the favourable process as it cannot be stored. (4) It is available based on geographies, such as land, lake, sea, mountains and the hills or plains.

Europe is the global pathfinder in the wind energy market and witness for the globalization of wind renewable energy. Many European countries such as Denmark, Germany, Belgium, France, Ireland, the Netherlands, Scotland, Sweden and the United

Kingdom are projected to generate a huge amount of renewable energy on the coasts of each country. Denmark is the world's largest manufacturer of wind turbines with almost 60% of the world's wind turbines manufactured there [11]. Germany, Spain and other European Union countries have made substantial progress in installing wind power plants for generating electricity. The goal set by the United States is that 20%–30% of total energy consumption will come from renewable energy resources.

Hydrogen is considered as one of the best alternative fuels as its oxidised product is water, which is an environmentally benign and lightweight material. Several nanomaterials have shown their advancement to efficiently convert water to H_2 either electrochemically or by using sunlight. Microbial and other biological pathways have recently been studied for the production of H_2 [8]. However, hydrogen storage is a severe problem as this field needs more time for industrial development and advancement. Researchers are currently focused on developing nanomaterials, which will have the storage property of this small fuel molecule as well as the conversion efficiency of this molecule to water whenever needed.

To achieve the best results out of alternative energy technology, improvements in storage technology need to be implemented. Alternative storage technologies such as metal-air battery, metal-ion battery, supercapacitor and hybrid supercapacitors are being seriously considered. Hybrid power supply systems, such as a fuel cell connected with another power supply system like supercapacitors, or secondary batteries, are revolutionising the alternative energy storage concept [14]. The advantage of using a supercapacitor in the hybrid system is its short charge-discharge time. Hybrid storage devices can in principle store power from different energy harvesters based on the mechanical, piezoelectric, triboelectric, pyroelectric, thermoelectric and photovoltaic effects. This type of hybrid device can be used to replace the usual battery or at least to extend the lifetime of the batteries.

III. SCOPE OF BOOK

Nanotechnology has shown advancement in the development of various renewable technologies. A significant improvement is seen in the development of solar cells and the generation of solar fuels with the help of various advanced materials. Nanomaterials including transition metal, metal oxide and nanocarbon could help solve the obstacles and meet clean energy expectations without disturbing the environment and human health.

Carbon is one of the most abundant elements on earth and has several allotropes that exist in nanoscale. Nano forms of carbon such as graphene, carbon nanotubes, fullerene, carbon nanofiber, activated carbon and several other forms have been intensively investigated in the past two decades in the field of clean, renewable and alternative energy technology. Owing to their unique properties such as high surface area, mechanical strength, electrical and thermal conductivity, different nanocarbon allotropes have opened new possibilities for their use in a wide range of applications. In this book, we emphasise different forms of renewable energy as well as the role of various nanocarbons and their properties in formulating renewable energy devices.

REFERENCES

1. G. Gilbert, World Population: A Reference Handbook, ABC-CLIO, Santa Barbara, CA, 2005.
2. R.F. Service, Solar energy: Is it time to shoot for the sun? *Science* 309, 2005, 548–551. doi: 10.1126/science.309.5734.548
3. M.A. Laughton, *Renewable Energy Sources* (Watt Committee Report; no. 22), 1990.
4. M. Nakamura, Nanotechnology for sustainable society, in: *Ext. Abstr. 2010 Int. Conf. Solid State Devices Mater.*, The Japan Society of Applied Physics, 2015, pp. 2–3. doi: 10.7567/ssdm.2010.pl-1-1
5. M. Notarianni, J. Liu, K. Vernon, N. Motta, Synthesis and applications of carbon nanomaterials for energy generation and storage, *Beilstein J. Nanotechnol.* 7, 2016, 149–196. doi: 10.3762/bjnano.7.17
6. B.P. Heard, B.W. Brook, T.M.L. Wigley, C.J.A. Bradshaw, Burden of proof: A comprehensive review of the feasibility of 100% renewable-electricity systems, *Renew. Sustain. Energy Rev.* 76, 2017, 1122–1133. doi: 10.1016/j.rser.2017.03.114
7. T.W. Brown, T. Bischof-Niemz, K. Blok, C. Breyer, H. Lund, B.V. Mathiesen, Response to 'Burden of proof: A comprehensive review of the feasibility of 100% renewable-electricity systems', *Renew. Sustain. Energy Rev.* 92, 2018, 834–847. doi: 10.1016/j.rser.2018.04.113
8. M.S. Dresselhaus, I.L. Thomas, Alternative energy technologies, *Nature* 414, 2001, 332–337. doi: 10.1038/35104599
9. D.J. Coyle, H.A. Blaydes, R.S. Northey, J.E. Pickett, K.R. Nagarkar, R. Zhao, J.O. Gardner, Life prediction for CIGS solar modules part 2: Degradation kinetics, accelerated testing, and encapsulant effects, *Prog. Photovoltaics Res. Appl.* 21, 2013, 173–186. doi: 10.1002/pip.1171
10. R.R. King, Record breakers, *Nat. Photonics.* 2, 2008, 284–286. doi: 10.1038/nphoton.2008.68
11. G.M. Joselin Herbert, S. Iniyan, E. Sreevalsan, S. Rajapandian, A review of wind energy technologies, *Renew. Sustain. Energy Rev.* 11, 2007, 1117–1145. doi: 10.1016/j.rser.2005.08.004
12. M. Ramesh, T.R. Jyothsna, A concise review on different aspects of wind energy system, in: *2016 3rd Int. Conf. Electr. Energy Syst. ICEES 2016, IEEE*, 2016, pp. 222–227. doi: 10.1109/ICEES.2016.7510644
13. R.S. Amano, Review of wind turbine research in 21st century, *J. Energy Resour. Technol.* 139, 2017, 050801. doi: 10.1115/1.4037757
14. T. Momma, H. Nara, T. Osaka, *Encyclopedia of Electrochemical Power Sources*, Academic Press, Amsterdam, 2009.

1 Clean and Renewable Energy Sources
Global Comparison

1.1 INTRODUCTION

Global climate change (global warming or the greenhouse effect) is the most significant environmental issue relating to energy. The momentousness of climate change originating from human activity is a global concern due to the release of greenhouse gasses. Many reports claim that the average surface temperature has increased by 0.4°C–0.8°C globally in the last century above the baseline of 14°C [1]. The upswing of earth's surface temperature, increasing the seawater level, acidification of the ocean and changes in weather patterns are the results [2,3]. The upsurge in the atmospheric carbon dioxide (CO_2) level occurs largely from the burning of fossil fuels and changing land use, such as deforestation. However, other greenhouse gasses such as methane (CH_4) and nitrous oxide (N_2O) increase its atmospheric concentrations, and mostly come from agriculture [4]. It is projected that greenhouse gas emission will be almost double from 13.5 TW (terawatt) at the start of the century to 28 TW by the middle of the century [5,6].

World population growth has always been a key driver for global energy demands. Table 1.1 presents the values for a number of indicators recorded in 1993 and also in 2011 and the predictions for 2020 by the Energy for Tomorrow's World, High-Growth Scenario A to 2020 [7]. As can be seen from Table 1.1, population growth has increased from 5.5 billion in 1993 to 7 billion in 2011 and is expected to be 8.1 billion in 2020. The overall demand for energy is becoming higher and higher due to population growth and average living standards.

According to the Inter-governmental Panel on Climate Change projection, it is reported that the projected carbon intensity in 2050 will be ~0.45 kg of carbon yr^{-1} W^{-1}, which is lower than that of any of the fossil fuels [6]. A significant contribution of carbon-free power to total energy is the only possible way to come close to this value of the mean carbon intensity. The role of every country is to set a roadmap towards obtaining 100% clean, sustainable and zero-energy emission technology. It was predicted that the 1.5°C warming could be avoided by converting the energy infrastructure to zero-emitting energy by no less than 80% by 2030 and 100% by 2050 [4]. A number of renewable and non-renewable energy sources are summarised in Figure 1.1. This chapter demonstrates the various forms of energy sources from non-renewable as well as renewable energy and their advantages and disadvantages in the future energy perspective.

TABLE 1.1
Key Indicators for the Years 1993, 2011 and 2020

	1993	2011	2020	Percentage (%) Growth 1993–2011
Population, billion	5.5	7	8.1	27
GDP (Trillion USD)	25	70	65	180
Coal (Mt)	4474	7520	10,108	68
Oil (Mt)	3197	3973	4594	25
Natural gas (bcm)	2176	3518	4049	62
Nuclear (TW h)	2106	2386	3761	13
Hydro power (TW h)	2286	2767	3826	21
Biomass (M ton)	1036	1277	1323	23
Other renewables (TW h)	44	515	1999	n/a
Electricity production/year (TW h)	12,607	22,202	23,000	76
CO_2 emission/year (total CO_2 Gt)	21	30	42	44
Energy intensity (koe; 2005 USD)	0.24	0.19	n/a	−21

Note: Collected from Tomorrow's World (World Energy Council [WEC], 1995) World Energy Resources (WEC, 2013) and World Energy Scenarios Report (WEC, 2013).

1.2 NON-RENEWABLE ENERGY RESOURCES

It is obvious that coal, oil and gases will be the key energy resources for the next few decades. In the last two decades, main energy sources have shifted from wood to coal to oil and currently to natural gases [2]. Consumption of fossil energy at that rate, however, will produce a potentially significant global issue. Important to future necessary growth is improvement in new technologies that focus on generating oil and gas from eccentric reservoirs in an environmentally suitable approach.

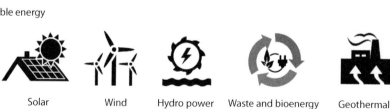

FIGURE 1.1 Various non-renewable and renewable energy sources.

1.2.1 COAL

As energy is an important prerequisite for modern life, coal is playing a significant position in providing energy access, because it is usually available, benign, reliable and relatively inexpensive. The top five coal reserves are located in the United States, the Russian Federation, China, Australia and India. Coal resources are easily available in many developing countries, particularly in Asia and Southern Africa. Coal is playing a key role in these countries as an inexpensive and safe way to fuel the growing need for electricity. Electricity is generated from the coal supply and sent to the domestic grids that distribute the energy to millions, hence accelerating economic progress in the developing world. However, coal-based electricity produces high emissions of CO_2, particulates and many pollutants.

1.2.2 OIL

Oil is an established global industry that gives the market contributors opportunities for lucrative revenues. The main use of oil is shifting towards the transport sector; however, in the near future, the oil sector is going to face challenge from other fuels. The top five oil reserve countries include Venezuela, Saudi Arabia, Canada, Iran and Iraq. Oil as a future energy source has several drawbacks, such as high price volatility and geographical and political tension in areas that have the greatest reserves.

1.2.3 NATURAL GAS

Natural gas is another important fossil fuel that significantly contributes to the global energy economy due to its availability and flexibility. It is progressively used in the highest-proficiency power generation technologies, such as the combined cycle gas turbine (CCGT), where the conversion efficiency is about 60%. The survey, development and transport of natural gas usually demands noteworthy open investment. Several shortcomings of using natural gas as a future energy source include high upfront investment for transport and distribution costs, high-cost infrastructures and offshore fields.

1.3 RENEWABLE ENERGY SOURCES

Renewable energy sources are projected to play an important role in the world's energy future. There are plenteous fossil energy reserves, in one form or another, which may supply the future energy necessity at some sensible cost. By definition, renewable energy is the energy that can be generated and continually replenished through natural processes, such as solar energy, wind energy, biomass-based energy, geothermal energy, etc. The employment of renewable energy is a viable, environmentally sensible solution to the rising large-scale demand for global energy. The key features of renewable energy technologies include substantial reductions in carbon footprint and environmental pollution. A brief discussion of such energy technologies including conventional photovoltaics, solar and thermal technologies, as well as passive solar technologies such as biofuels, biomass and wind power, is presented here, and detailed discussion can be found in the following chapters.

1.3.1 SOLAR ENERGY

Solar energy is the best abundant energy resource, which creates a usable impact with almost zero environmental impact. There are two types of solar energy forms: direct (solar radiation) and indirect (wind, biomass, hydro, ocean, etc.). It is well known that approximately 60% of the total energy emitted by the sun reaches the earth's surface. However, only 0.1% of this energy could be efficiently converted at an efficiency of 10% [8]. The generation of solar energy requires the following three steps: (1) capture of light, (2) conversion of solar light and (3) storage of converted energy. Solar photovoltaic (PV) cells are usually made with semiconductor material to produce a stream of electricity when coming into contact with sunlight [9]. The total magnitude of solar energy reaching the earth's surface per day from the sun is huge, but the available energy at any specific point is difficult to measure. The sum of energy resulting from solar panels is subject to the ambient solar level as well as the collector's energy transformation efficiency.

It should be pointed out here that although the technology is now successfully industrialised and trustworthy, it is costly compared to current energy sources, perhaps three times as expensive as fossil-fuel engendered electricity [10]. The global challenge here is to drastically reduce the cost per Watt of supplied solar electricity.

1.3.2 WIND ENERGY

Wind energy is a popular source of renewable energy as it is available virtually everywhere on earth. Wind energy is the energy generated by moving air, and the obtainable energy varies with the cube of wind velocity. Even though wind power is most often used to power small populations situated close to wind farms, it is becoming a progressively more popular energy source and could presently power a whole city. The global capacity of wind energy has been doubling about every three and a half years since 1990. Denmark is using wind power to produce approximately 25%–30% of its national electricity system, which is the highest in the world [7,11]. A few shortcomings of the wind power technologies are intermittency, grid integration challenges and reliance on subsidies. To encounter future demand, many wind turbines will be positioned offshore, exceed 3 MW maximum rated output and be expected to have lower operating and maintenance costs, be more consistent, and have a bigger local production [12].

1.3.3 WASTE AND BIOENERGY

Bioenergy/waste energy is a different kind of energy that is generated from a variety of feedstocks of biological sources and by copious conversion technologies to generate different forms of energy such as heat, power, liquid biofuels and gaseous biofuels [2]. Various industrial raw materials such as pulp, paper, tobacco, and so on, generate by-products such as bark, wood chips, black liquor and agricultural residues, which are then converted to bioenergy and biofuel.

Biofuel or biomaterials are carbon-negative or zero-carbon technologies that can decrease the atmospheric concentration of carbon dioxide by converting it into

Clean and Renewable Energy Sources

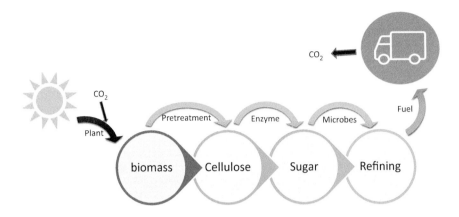

FIGURE 1.2 Process of conversion from biomass to biofuel.

worthwhile materials. It was recently reported that biofuels are one of the least expensive technologies to implement at the gigaton scale (the scale required to decrease CO_2 equivalent emissions by 1 Gt per year) [13]. Biofuel can be synthesised using a bioreactor in which microorganisms transform the cellulose into sugars that can be metabolised (by microorganisms) to produce fuels, as shown in Figure 1.2. It is a massive challenge to synthesise biofuel at the scale and cost required to substitute for petroleum, given economical and land-use considerations. In spite of the advantage of being domestically available and using simple conversion technologies, it has a few drawbacks such as transportation and processing effects and NO_x and SO_x emission during their conversion.

1.4 ROLE OF CARBON NANOMATERIALS

In the last two decades, carbon, one of the most abundant materials found on earth, and its allotrope in different forms such as fullerenes, carbon nanotubes and graphene are being proposed as sources of energy generation and storage due to their extraordinary properties and ease of production. For the synthesis of carbon nanomaterials, various methodologies have been incorporated in photovoltaics, fuel cells and supercapacitors. In Chapter 2, the synthesis of various allotropes is discussed in detail.

1.5 CONCLUSION

It is obvious from the discussion in this chapter that fuel-switching does not happen overnight. Moving away from fossil fuels will take years and decades, as coal, oil and gas will remain the main energy resources in many countries. The potential for growth in the use of renewable resources such as, in particular, wind power and solar PV is comparatively better than that for other forms of renewable energy. Nanomaterials, mainly nanocarbon allotropes, can be used as fairy lights for future generations. As carbon nanomaterials are inexpensive to synthesise and they possess extraordinary properties, they can be used as efficient materials for sustainable energy development.

The energy industries are looking for cost-effective suitable nanomaterials that can supply fuels to meet global energy demands. Necessary improvements in storage and battery materials must be made in order for these to be commercially viable.

REFERENCES

1. N.L. Panwar, S.C. Kaushik, S. Kothari, Role of renewable energy sources in environmental protection: A review, *Renew. Sustain. Energy Rev.* 15, 2011, 1513–1524. doi: 10.1016/j.rser.2010.11.037
2. D. S. Ginley, D. Cahen, editors, *Fundamentals of Materials for Energy and Environmental Sustainability*, Cambridge University Press, Cambridge, UK, 2012.
3. B. Obama, The irreversible momentum of clean energy, *Science* 355, 2017, 126–129. doi: 10.1126/science.aam6284
4. M.Z. Jacobson, M.A. Delucchi, Z.A.F. Bauer, C. Savannah, W.E. Chapman, M.A. Cameron, A. Cedric et al., 100% Clean and renewable Wind, Water, and Sunlight (WWS) all-sector energy roadmaps for 139 countries of the world by 2050, *Draft 1*, 2015, 1–61.
5. R.E. Hester, R.M. Harrison, *Sustainability and Environmental Impact of Renewable Energy Sources*, Royal Society of Chemistry, Cambridge, UK, 2007. doi: 10.1039/978 1847551986
6. N.S. Lewis, D.G. Nocera, Powering the planet: Chemical challenges in solar energy utilization, *Proc. Natl. Acad. Sci.* 103, 2006, 15729–15735. doi: 10.1073/pnas.0603395103
7. World Energy Council, World Energy Resources: 2013 survey, World Energy Council, 2013, 11. http://www.worldenergy.org/wp-content/uploads/2013/09/Complete_WER_2013_Survey.pdf
8. H.P. Garg, S.C. Mullick, A.K. Bhargava, *Solar Thermal Energy Storage*, Springer Netherlands, Dordrecht, 1985. doi: 10.1007/978-94-009-5301-7
9. B.J.M. de Vries, D.P. van Vuuren, M.M. Hoogwijk, Renewable energy sources: Their global potential for the first-half of the 21st century at a global level: An integrated approach, *Energy Policy* 35, 2007, 2590–2610. doi: 10.1016/j.enpol.2006.09.002
10. D. Timmons, J.M. Harris, B. Roach, The economics of renewable energy, *Renew. Energy* 15081, 2009, 1341–1356. doi: 10.3386/w15081
11. F. Kreith, Y. Goswami, *Handbook of Energy Efficiency and Renewable Energy*, Choice Rev. Online. 45, 2008, 45-2629-45-2629. doi: 10.5860/CHOICE.45-2629
12. R.E.H. Sims, H.H. Rogner, K. Gregory, Carbon emission and mitigation cost comparisons between fossil fuel, nuclear and renewable energy resources for electricity generation, *Energy Policy* 31, 2003, 1315–1326. doi: 10.1016/S0301-4215(02)00192-1
13. The Gigaton Throwdown Initiative, http://www.gigatonthrowdown.org. (n.d.)
14. A. Grubler, M. Jefferson and, N. Nakićenović, Global Energy Perspectives: A Summary Of The Joint Study By IIASA And World Energy Council, *Technological Forecasting and Social Change* 51, 1996, 237–264.

2 Different Allotropes of Carbon, Their Structures and Properties

2.1 INTRODUCTION AND GENERAL INFORMATION

The name *carbon* came from the Latin word 'Carbo', which means 'charcoal' or 'ember' [1]. The chemical versatility and extraordinary capability of chemical carbon gives rise to the different forms of solid carbon. Carbon exists in various forms – crystalline, amorphous and nanoparticles – as well as engineered disordered structures. After the discovery of fullerene in 1985, there has been a surge of interest in carbon nanomaterial research.

Carbon is, indeed, a strange element, because its properties and applications are broad and varied. The electronic configuration of carbon is $1s^2\ 2s^2\ 2p^2$, which allows it to form three different kinds of bonds – single, double and triple. Figure 2.1 highlights the various hybridisation modes of carbon and their different allotropes. The usefulness of carbon comes from the fact that it can bond with other atoms, based on hybridisation with 2s and 2p atomic orbitals in three different ways: single bonding sp^3 (tetrahedral), double bonding sp^2 (trigonal planar) and triple bonding sp (linear). Therefore, depending on the hybridisation and structures, the carbon family tree covers graphite, diamond and amorphous carbons, nano-form carbon such as fullerenes, carbon nanotubes and graphene.

Carbon-based nanotechnology has unlocked a new frontier in materials science and engineering to meet various challenges by creating new materials. Therefore, carbon nanomaterials are introduced as a catalyst for efficient electrochemical energy conversion and storage. One can consider carbon for conventional energy materials, as nanomaterials such as carbon possess unique size- and surface-dependent properties (e.g. morphological, electrical, optical and mechanical) that are worthwhile for enhancing energy-conversion and storage performance. Carbon nanomaterials have been used over the last few decades in different applications to make the most of the unique properties of carbon.

2.2 THE ELEMENT CARBON AND ITS ALLOTROPES

The element carbon can exist in different forms. Figure 2.1 shows that diamond, graphite and amorphous carbon are the main allotropes in the carbon family tree. The sp^3 hybridisation of diamond makes it the hardest natural material, whereas graphite is a sp^2 hybridized solid material owing to the loose interlamellar π–π coupling present between the sheets. In the past three decades, many new carbon nanomaterials were

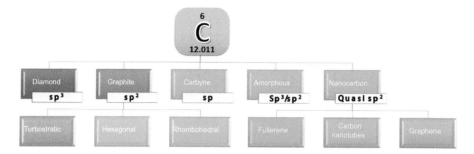

FIGURE 2.1 Different allotropes, types and hybridisation of carbon.

discovered. Experimental findings together with theoretical calculations illustrate the outstanding physicochemical properties of many of these, which led to many new carbon nanoforms being investigated. Several forms of carbon allotropes including fullerene, CNT, graphene, and so on are on the list of recent findings. Among them, it is considered that graphene is a basic structure for all graphitic materials for all other dimensionalities [2]. The various carbonaceous materials with different dimensionalities are listed in Figure 2.2.

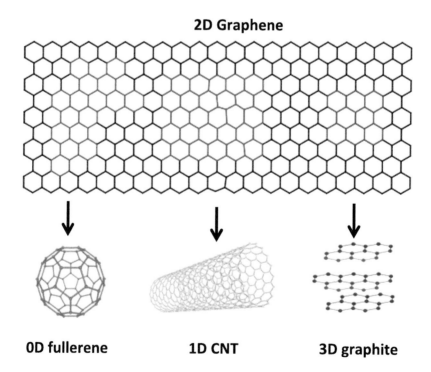

FIGURE 2.2 Different dimensions of nanocarbon 0D fullerene, 1D carbon nanotube and 3D graphite structures can be derived from 2D graphene.

2.2.1 GRAPHITE: MOST USED CARBON ALLOTROPES IN ENERGY

Graphite has been widely studied, and its structure and properties are well described in the literature. Graphite has a planar layered structure with separation of layers of 0.335 nm. The carbon atom in graphite is sp^2 hybridised. The fourth electron of each carbon atom forms delocalized π-bonds that spread equally over all carbon atoms as can be seen in Figure 2.3. Thanks to these π-electrons, graphite conducts electricity parallel to the planes. Layers of graphite slide over each other effortlessly because only weak forces exist between the layers, making graphite slippery. Due to the high electrical conductivity of graphite, it has been extensively used as an electrode for different energy applications. Graphite has been extensively used as anodal electrode material in lithium ion batteries (LIBs) [3]. It is a well-studied material and can be optimised for energy and power applications. Graphite is the material of choice for the electrode required for fuel cell applications.

2.2.2 GRAPHENE AND REDUCED GRAPHENE OXIDE

Graphene, a two-dimensional (2D) wonder material, is considered as a basic building block for all graphitic material [2,4]. Graphene is an extended honeycomb network of sp^2 hybridised carbon that was discovered in 2004 by Andre Geim and Konstantin Novoselov, who subsequently were awarded the Nobel Prize in physics in 2010. Graphene can be wrapped up to form 0D fullerenes, rolled to form 1D carbon nanotubes, and stacked to form 3D graphite, as can be seen in Figure 2.2.

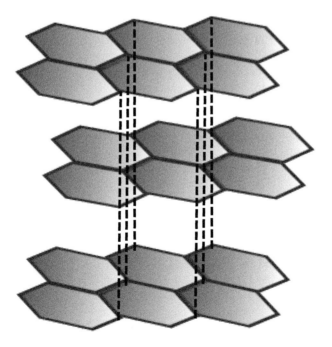

FIGURE 2.3 The π-bonds between the hexagonal sheets of carbon atoms.

The experimental discovery of single-layer graphene most importantly generated interesting physics knowledge. Graphene has been extensively studied so far mainly due to its excellent properties. The extraordinary properties of graphene include very high electrical and thermal conductivity, anomalous integer quantum Hall effect at room temperature, mechanical stiffness, strength and elasticity, and many others. In recent years, graphene research not only has been subjected to one-atomic single-layer graphene alone but also to bi- and few-layer (<10 layers) graphene [5].

The unit cell of single-layer graphene contains two atoms. Single-layer graphene can be defined as a 2D hexagonal sheet of carbon atoms. However, bi-, tri- and few-layer graphenes have 2 and 3 to 10 layers of such 2D hexagonal sheets, respectively. If the graphene consists of more than 10 layers, it is considered as a thick graphene sheet and has less scientific interest [6]. One can have a closer look at the structure of graphene, where C atoms in bi- and few-layer graphene can be stacked in different ways, generating hexagonal AA... stacking, Bernal AB... stacking and rhombohedral ABC... stacking [7]. The electronic structure of single-layer graphene shows band overlap in two conical points (K and K′) in the Brillouin zone. Hopping between the two equivalent carbon sublattices A and B leads to the π bands intersecting at the zone boundary, 'K-point', which are known as Dirac points. The honeycomb lattice of graphene is described by the (2 + 1)-dimensional Dirac equation at low E, with an effective speed of light, $v_F = 10^6$ m s^{-1} [7–9].

The current state of graphene synthesis is split into two approaches – the top-down and the bottom-up approaches. The top-down synthesis of graphene is governed by the fundamental idea of removing layers of graphene from graphite by various methods. However, the bottom-up methodology is discovered to produce graphene sheets starting with simple carbon molecules such as methane and ethanol. If the intent is to incorporate graphene into batch-type synthesis, solution-based chemistry has to be developed for forming blends and graphene composites.

As the graphene sheets are 2D in nature, the edge areas have an important role in the electronic structure of the molecules. Figure 2.4 describes the edges of graphene, noticeable as either *zigzag* tracks or *armchair* tracks. In the *zigzag* edges, the understanding of a six-membered ring is lost in the majority of the rings; hence such a structure is thermodynamically unbalanced compared to the *armchair* edges. It can be anticipated that the *zigzag* edges will display higher reactivity as compared to the *armchair* edges. Parallel understanding of graphene reactivity can be obtained from graphene-like polyaromatic hydrocarbons where stable zigzag-edged molecules are more challenging to synthesise than *armchair* molecules.

2.2.3 Three-Dimensional Reduced Graphene Oxide

Graphene or 2D reduced graphene oxide and its functionalised derivatives have been used as the building block for self-assembled or template-directed organisation of the 3D architecture of reduced graphene oxide [10]. The driving force of forming 3D structures is through strong covalent bonding forces, weak π–π interactions, or even hydrogen bonding [11]. The 3D architecture can largely maintain the unique properties of discrete graphene sheets and has recently drawn extreme interest for fundamental investigations and potential applications in miscellaneous

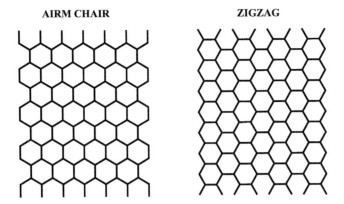

FIGURE 2.4 Zigzag and armchair edges of a graphene sheet.

technologies [12]. The 3D graphene (3DG)–based composites are used in a wide range of energy applications, for example, supercapacitors, lithium-ion batteries, dye-sensitized solar cells and fuel cells; thus it is necessary to gain a better understanding and further improve 3DG-based composites in these energy systems/devices. Multiple reaction approaches, comprising hydrothermal/solvothermal reduction, chemical reduction and electrochemical reduction, have been established for the preparation of 3D graphene macrostructures [13].

2.2.4 Carbon Nanotubes

Carbon nanotube (CNTs) are theoretically unique as a cylinder-like structure of a rolled-up graphene sheet. CNTs are of two types – single wall or multiple walls. CNTs with a single wall are labelled as single-walled carbon nanotubes (SWCNTs) and were first reported in 1993 [14], whereas CNTs with multiple walls are described as multi-walled carbon nanotubes (MWCNTs) [15]. Almost all physical properties of carbon nanotubes are expressed from graphene. Carbon atoms are densely organised in a regular sp^2-bonded atomic-scale honeycomb (hexagonal) pattern, and this pattern is a basic structure for CNTs, similar to graphene [16].

The diameter and helicity of a SWCNT are uniquely characterised by the chiral roll-up vector $\mathbf{C} = n\mathbf{a}_1 + m\mathbf{a}_2$ ($\mathbf{a}_1$ and $\mathbf{a}_2$ are the lattice vectors of graphite, m and n are the integers), and this vector determines the direction of rolling a graphene sheet [16,17]. Depending on the chiral indices (n, m), CNTs can be classified as zigzag and armchair structures, as can be observed in Figure 2.5. The chiral roll-up vector is denoted by a pair of indices, n and m, where these two integers resemble the number of unit vectors along the two directions in the honeycomb crystal lattice of graphene. When m = 0 the nanotube is called 'zigzag', when n = m the nanotube is called 'armchair'.

Carbon nanotubes hold extraordinary electronic properties such as high electrical conductivity (comparable to metal). It is interesting to note that nanotubes can be metallic or semiconducting depending on their electronic band structure. When the rolling action of the plane breaks the symmetry, a different direction is imposed

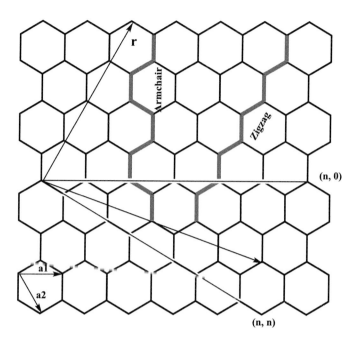

FIGURE 2.5 Vectors indicating rolling of the graphene sheet responsible for formation of zigzag and armchair edges.

with respect to the hexagonal lattice and the axial direction. Carbon nanotubes may perhaps perform electrically as either a metal or a semiconductor depending on the connection between this axial direction and the unit vectors defining the hexagonal lattice. A semiconducting carbon nanotube has bandgaps that scale inversely with diameter, ranging from 1.8 eV (for very small diameter tubes) to 0.18 eV (for the widest possible stable SWCNT) [18–20].

2.2.5 Fullerene

In 1985, a new allotrope of carbon was discovered by Harold W. Kroto, Robert F. Curl and Richard E. Smalley and was named Buckminster fullerene, also called the Bucky ball [21]. Interestingly, the truncated icosahedrons fullerene C_{60} is the easiest to produce and thus is most popularly applied in a variety of fields. In a very unusual design, they found some uncommon symmetrical structures of carbon by laser vaporisation of graphite and named it Buckminster fullerene (C_{60}), after an architect-engineer, R. Buckminster Fuller. After the first discovery of Buckminster fullerene (C_{60}), various types of fullerene such as, C_{70}, C_{76}, C_{78}, C_{80}, C_{82} and C_{84} were discovered. The cage-like structure of a fullerene consists of 12 five-member rings and a number of six-member rings depending on the number of carbon atoms [22]. It was postulated that the five-member rings need to be separated from each other to reduce the localisation of the strain caused by the bending of the sp^2-hybridised carbon atoms. This is the gist of the isolated pentagon rule (IPR) [23] which dramatically reduces the possible number of isomers for a given fullerene family. The major isomers of fullerenes are C_{60} to C_{84}. Any closed convex polyhedron assembled from polygons with equal-length sides must satisfy Euler's polyhedron formula. This relates the number of vertices (V), edges (E) and faces (F) for a convex polyhedron with equal-length sides as $V - E + F = 2$.

The C_{60} molecule is known for its symmetry. It has three types of rotation axes, where the fivefold rotational symmetry is the most common, although it possesses twofold and threefold rotational symmetry axes. Hereafter, it is called the *maximum symmetric molecule*. Fullerenes are expressed by the general chemical formula $C_{20} + 2H$, where H is the number of hexagonal faces. Fullerenes are more similar to graphite with respect to the bond distance between the carbon–carbon atoms in fullerene. The C_{60} fullerene has been synthesised from the natural plant-based precursor camphor by Sharon et al. The C_{60} fullerene molecule has a system of mobile electrons which donate additional stabilisation to the polyhedron as in the benzene molecule. It formed the first quantifiable assessments of resonance energy for this unusual type of conjugated molecule and established that the C_{60} cluster has greater resonance energy than the benzene molecule. Hence, C_{60} fullerene is consigned to a new course of aromatic system.

2.2.6 Activated Carbon

Activated carbon is associated with a family of carbons ranging from carbon blacks to nuclear graphite, from carbon fibers and composites to electrode graphite, and many more. Activated carbon is one of the most useful bio-based materials known to human beings. A naturally obtainable activated carbon is charcoal, which results from burning wood [24]. The earliest uses of charcoal in history were for fuel, wood preservation and medicine for a number remedies such as for treating gangrenous ulcers, epilepsy, intestinal disorder and chlorosis [25]. Only a few resources are proven as useful for activated carbon production, including coals of several rank, peat as well as biomass such as woods, fruit stones and nutshells, as with peanut shells, coconut shells along with some synthetic organic polymers. An essential element is the consistency and reliability of the source. The built-up procedures are exceptionally tuned and variations in the superiority of the resource are intolerable. There are other resources that frequently gain attention, many biomass and waste materials such as banana skins, human hair, straw, woodcuttings, fruit shells, casings from coffee beans and many other organic waste materials.

Activated carbon has a structure like non-graphitizing carbon that is related to that of the fullerenes. It consists of curved fragments containing pentagons and other non-hexagonal rings in addition to hexagons [26]. This type of structure would explain the microporosity of the carbon, and many of its other properties, which make it suitable for energy applications.

2.2.7 Carbon Nanofiber

Carbon nanofiber (CNF) is demonstrated as a fiber containing at least 92 wt% carbon; however, the fiber having at least 99 wt% carbon is generally denoted as a graphite fiber [27]. In 1879, during his work on the incandescent light bulb, Thomas Edison discovered that carbon fiber filaments were found when baking cotton threads or bamboo strips [28]. Akio Shindo in Japan at the same time pursued research on heat-treated polyacrylonitrile (PAN) fibers, resulting in PAN-based carbon fibers with tensile modulus values as high as 140 GPa [29]. Carbon nanofibers usually ensure good tensile properties, high thermal and chemical stability in the absence of oxidising agents, low densities, and good thermal and electrical conductivities. The present carbon fiber market is governed by PAN carbon fibers, and they have been used in composites in the form of woven textiles, prepregs, and continuous and chopped fibers [27].

The structure of a carbon fiber is similar to that of graphite, consisting of carbon atom layers like graphene sheets arranged in a regular hexagonal pattern. The layered planes in carbon fibers may be turbostratic or graphitic or form a hybrid structure. One-dimensional (1D) nanostructures such as CNF can exhibit strong mechanical properties that are essential for atomic-scale manipulation and advanced modifications. Because of the large surface area and electronegative absorption, 1D nanomaterials are proficient in interacting with various inorganic species, which is suitable for electrode materials. Recently, electrospun CNF has had a strong push towards the development of clean and efficient energy systems for both conversion (such as solar cells and fuel cells) and storage (supercapacitors, batteries, particularly lithium-ion and hydrogen storage) [30–32].

2.2.8 OTHER FORMS OF CARBON

Nano dimensions of carbon have several other forms such as carbon nano-onion, carbon nanocones, carbon nanobelts, etc.

2.2.8.1 Carbon Nano-Onions

The carbon nano-onion (CNO) was first discovered by Sumio Iijima in 1980 while looking at a sample of carbon black in a transmission electron microscope. CNO was not produced in bulk, but rather was observed as a by-product of carbon black synthesis [33]. CNO is one of the carbon allotropes, composed of multilayered concentric graphitic shells. Carbon onions or CNOs contain spherical closed carbon shells, and its name is due to the concentric layered structure resembling that of an onion [34]. Carbon nano-onions are sometimes called onion-like carbon (OLC). CNOs do not possess a porous structure due to their closed, concentric graphitic helices and most of the preparation methods have a low degree of graphitisation. A CNO exhibits excellent performance as an anode in a LIB, because it has an anisotropic surface due to the quasi-spherical structure in combination with considerable surface area (300–600 $m^2\ g^{-1}$) resulting in a considerable contact area between the active material and the electrolyte [35].

2.2.8.2 Carbon Nanocones

A carbon nanocone (CNC) is a closed cage of sp^2-bonded carbon atoms with a diameter of 2–5 nm and length of 40–50 nm. CNCs can be considered a high-aspect ratio subclass of fullerenes owing to their closed cage structure [36]. The carbon cones can be demonstrated as being composed of curved graphite sheets formed as open cones. Figure 2.6 shows the formation of the CNCs, which can be imagined by cutting out the sectors of $n \times 60°$ (n = 1 − 5) from the flat sheet of graphene

Different Allotropes of Carbon, Their Structures and Properties

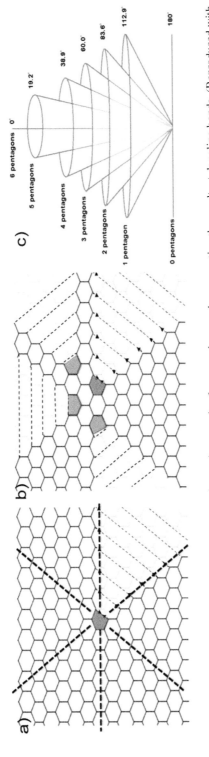

FIGURE 2.6 Construction of carbon nanocones by cutting a wedge from graphene and reconnecting the resultant dangling bonds. (Reproduced with permission from N. Karousis et al., *Chem. Rev.*, 116, 2016, 4850–4883.)

and subsequently connecting the edges [37]. Nanocone helicity is not constant and increases monotonously along the cone axis; hence, both chiral and achiral nanocones are possible. CNCs are synthesised mostly by Arc discharge methods, although there are few reports via Joule heating. CNCs are well adapted to large-scale production due to their potential application as electrodes for fuel cells.

2.3 SUMMARY

Carbonaceous nanomaterials are becoming the focus of attention of physicists and chemists because of their distinctive properties of sp^2-hybridized carbon bonds with unusual characteristics at the nanoscale. The exceptional properties of carbonaceous nanomaterials commonly mentioned in energy applications are size, shape and surface area; molecular interactions and sorption properties; and electronic, optical and thermal properties. Carbon has several allotropes or various forms in nanoscale depending on its size and dimension, which continues its attraction to the nanoscientist. This chapter is an effort to summarise the structures of various allotropes of nanocarbon and the properties that have been discovered so far.

REFERENCES

1. H.O. Pierson, *Handbook of Carbon, Graphite, Diamond and Fullerenes Properties, Processing and Applications*, (n.d.).
2. A.K. Geim, K.S. Novoselov, The rise of graphene, *Nat. Mater.* 6, 2007, 183–191. doi: 10.1038/nmat1849
3. C. Engineering, *Carbon Nanomaterials*, CRC Press, Boca Raton, FL, 2006. doi: 10.1201/9781420009378
4. M.J. Allen, V.C. Tung, R.B. Kaner, Honeycomb carbon: A review of graphene, *Chem. Rev.* 110, 2010, 132–145. doi: 10.1021/cr900070d
5. C.N.R. Rao, A.K. Sood, R. Voggu, K.S. Subrahmanyam, Some novel attributes of graphene, *J. Phys. Chem. Lett.* 1, 2010, 572–580. doi: 10.1021/jz9004174
6. W. Choi, I. Lahiri, R. Seelaboyina, Y.S. Kang, Synthesis of graphene and its applications: A review, *Crit. Rev. Solid State Mater. Sci.* 35, 2010, 52–71. doi: 10.1080/10408430903505036
7. J. Hass, W.A. De Heer, E.H. Conrad, The growth and morphology of epitaxial multilayer graphene, *J. Phys. Condens. Mater.* 20, 2008. doi: 10.1088/0953-8984/20/32/323202
8. S. Dröscher, P. Roulleau, F. Molitor, P. Studerus, C. Stampfer, K. Ensslin, T. Ihn, Quantum capacitance and density of states of graphene, *Appl. Phys. Lett.* 96, 2010, 15–17. doi: 10.1063/1.3391670
9. F. Banhart, J. Kotakoski, A. V. Krasheninnikov, Structural defects in graphene, *ACS Nano.* 5, 2011, 26–41. doi: 10.1021/nn102598m
10. S. Nardecchia, D. Carriazo, M.L. Ferrer, M.C. Gutiérrez, F. del Monte, Three dimensional macroporous architectures and aerogels built of carbon nanotubes and/or graphene: Synthesis and applications, *Chem. Soc. Rev.* 42, 2013, 794–830. doi: 10.1039/c2cs35353a
11. A.-H. Lu, G. Hao, Q. Sun, Design of three-dimensional porous carbon materials: From static to dynamic skeletons, *Angew. Chemie Int. Ed.* 52, 2013, 7930–7932. doi: 10.1002/anie.201302369
12. T. Purkait, G. Singh, D. Kumar, M. Singh, R.S. Dey, High-performance flexible supercapacitors based on electrochemically tailored three-dimensional reduced graphene oxide networks, *Sci. Rep.* 8, 2018, 640. doi: 10.1038/s41598-017-18593-3

13. B. Qiu, M. Xing, J. Zhang, Recent advances in three-dimensional graphene based materials for catalysis applications, *Chem. Soc. Rev.* 47, 2018, 2165–2216. doi: 10.1039/c7cs00904f
14. S. Iijima, T. Ichihashi, Single-shell carbon nanotubes of 1 nm diameter, *Nature.* 363, 1993, 603–605. doi: 10.1038/363603a0
15. S. Iijima, Helical microtubules of graphitic carbon, *Nature.* 354, 1991, 56–58. doi: 10.1038/354056a0
16. A. Eatemadi, H. Daraee, H. Karimkhanloo, M. Kouhi, N. Zarghami, A. Akbarzadeh, M. Abasi, Y. Hanifehpour, S.W. Joo, Carbon nanotubes: Properties, synthesis, purification, and medical applications, *Nanoscale Res. Lett.* 9, 2014, 1–13. doi: 10.1186/1556-276X-9-393
17. T.W. Odom, J.-L. Huang, P. Kim, C.M. Lieber, Structure and electronic properties of carbon nanotubes, *J. Phys. Chem. B.* 104, 2000, 2794–2809. doi: 10.1021/jp993592k
18. V.N. Popov, Carbon nanotubes: Properties and application, *Mater. Sci. Eng. R Reports* 43, 2004, 61–102. doi: 10.1016/j.mser.2003.10.001
19. B.K. Kaushik, M.K. Majumder, *Carbon Nanotube Based VLSI Interconnects*, Springer India, New Delhi, 2015. http://link.springer.com/10.1007/978-81-322-2047-3.
20. M. Benhamou, Advances in carbon nanomaterials: Science and applications, by Nikos Tagmatarchis, *Contemp. Phys.* 54, 2013, 135–136. http://www.tandfonline.com/doi/abs/10.1080/00107514.2013.806594
21. H.W. Kroto, J.R. Heath, S.C. O'Brien, R.F. Curl, R.E. Smalley, C60: Buckminsterfullerene, *Nature* 318, 1985, 162.
22. P.P. Shanbogh, N.G. Sundaram, Fullerenes revisited, *Resonance* 20, 2015, 123–135. doi: 10.1007/s12045-015-0160-0
23. J. Yadav, Fullerene: Properties, synthesis and application *J. Phys.* 6, 2017, 1–6.
24. J.Y. Chen, *Activated Carbon Fiber and Textiles*, Elsevier, New York, NY, 2016, pp. 143–242. doi: 10.1016/C2014-0-03521-6.
25. H. Marsh, F. Rodríguez-Reinoso, Characterization of activated carbon, in: H. Marsh, F. Rodríguez-Reinoso, editors, *Activated Carbon*, Elsevier, New York, NY, 2006, pp. 143–242. https://linkinghub.elsevier.com/retrieve/pii/B9780080444635500182
26. P.J.F. Harris, Z. Liu, K. Suenaga, Imaging the atomic structure of activated carbon, *J. Phys. Condens. Matter* 20, 2008. doi: 10.1088/0953-8984/20/36/362201
27. X. Huang, Fabrication and properties of carbon fibers, *Materials (Basel)* 2, 2009, 2369–2403. doi: 10.3390/ma2042369
28. N. Stevenson, Electric lamp, *Br. Med. J.* 1, 1883, 435. doi: 10.1136/bmj.1.1157.435-b
29. B.A. Newcomb, Composites: Part A processing, structure, and properties of carbon fibers, *Compos. Part A.* 91, 2016, 262–282. doi: 10.1016/j.compositesa.2016.10.018
30. S. Peng, L. Li, J. Kong Yoong Lee, L. Tian, M. Srinivasan, S. Adams, S. Ramakrishna, Electrospun carbon nanofibers and their hybrid composites as advanced materials for energy conversion and storage, *Nano Energy* 22, 2016, 361–395. doi: 10.1016/j.nanoen.2016.02.001
31. B. Zhang, F. Kang, J.M. Tarascon, J.K. Kim, Recent advances in electrospun carbon nanofibers and their application in electrochemical energy storage, *Prog. Mater. Sci.* 76, 2016, 319–380. doi: 10.1016/j.pmatsci.2015.08.002
32. J.K. Chinthaginjala, K. Seshan, L. Lefferts, Preparation and application of carbon-nanofiber based microstructured materials as catalyst supports preparation and application of carbon-nanofiber based microstructured materials as catalyst supports, *Ind. Eng. Chem. Res.* 46, 2007, 3968–3978. doi: 10.1021/ie061394r
33. S. Iijima, Direct observation of the tetrahedral bonding in graphitized carbon black by high resolution electron microscopy, *J. Cryst. Growth* 50, 1980, 675–683. doi: 10.1016/0022-0248(80)90013-5

34. J.K. McDonough, Y. Gogotsi, Carbon onions: Synthesis and electrochemical applications, *Interface Mag.* 22, 2013, 61–66. doi: 10.1149/2.F05133if
35. Y. Zheng, P. Zhu, Carbon nano-onions: Large-scale preparation, functionalization and their application as anode material for rechargeable lithium ion batteries, *RSC Adv.* 6, 2016, 92285–92298. doi: 10.1039/C6RA19060J
36. N. Karousis, I. Suarez-Martinez, C.P. Ewels, N. Tagmatarchis, Structure, properties, functionalization, and applications of carbon nanohorns, *Chem. Rev.* 116, 2016, 4850–4883. doi: 10.1021/acs.chemrev.5b00611
37. S.N. Naess, A. Elgsaeter, G. Helgesen, K.D. Knudsen, Carbon nanocones: Wall structure and morphology, *Sci. Technol. Adv. Mater.* 10, 2009. doi: 10.1088/1468-6996/10/6/065002

3 Synthesis and Characterisation of Carbonaceous Materials

3.1 INTRODUCTION

Carbon is one of the most abundant elements on earth, and it exists in different forms according to their electronic structure and hybridisation. In parallel with graphite and diamond, the discovery of fullerenes and subsequently of carbon nanotubes (CNTs) and graphene added a new dimension in nanoscience and nanotechnology.

In recent years, graphene, a planar sheet of one-atom-thick sp^2-bonded carbon atoms in honeycomb lattice, has drawn extensive attention as a next-generation electronic material. Graphene has exceptional physical and chemical properties such as high thermal conductivity, ballistic transport, high current density, chemical inertness, optical transmittance, super hydrophobicity at the nanometre scale and enormously high specific surface area [1]. Electronic characterisation such as transport measurements has shown that graphene exhibits electron and hole mobility values of more than 200,000 cm^2 V^{-1} s^{-1} at room temperature. The resistivity of the graphene sheet is calculated to be 10^{-6} ohm–cm, which is less than the resistivity of silver, the lowest resistive substance known at room temperature [2–6]. Single-layer graphene can be obtained from mechanical exfoliation of natural graphite or highly oriented pyrolytic graphite (HOPG), but the process is time consuming, and yield is comparatively small. It is reported that by the annealing of SiC substrates, a monolayer or 'decoupled' multilayer graphene can be generated [7,8]. Reduced graphene oxide sheets can be produced by the chemical reduction of graphene oxide platelets, which show relatively low electrical conductivity and have been ascribed to structural defects that occur during oxidation and reduction processes [9–11].

CNTs exhibit a cylindrical nanostructure, composed entirely of sp^2-bonded carbon atoms, and present a seamless structure with hexagonal honeycomb lattices, being a few nanometers in diameter and several microns in length. They are generating interest in nanoscience and engineering because of their exceptional physical, mechanical, thermal and optical properties. Structurally, CNTs can be described as a sheet of graphene rolled into a tube. CNTs are believed to have potential applications as high-strength engineering fibers, catalyst supports in heterogeneous catalysis and molecular wires for the next generation of electronic devices [12]. Armchair CNTs (n = m) typically show metallic conductivity, whereas zigzag (m = 0) or chiral (n ≠ m) CNTs are semiconducting (Figure 2.5, Chapter 2). Different forms of CNTs exist depending on the number of carbon layers, such as single-walled CNT (SWCNT), double-walled CNT (DWCNT) and multi-walled CNT (MWCNT). Three common methods for synthesising SWCNTs and MWCNTs are reported including electrical arc discharge, laser ablation (vaporisation) and chemical vapor deposition (CVD) [13].

This chapter focuses on the various synthesis methods currently available for the fabrication of graphene and CNTs. The physical and chemical properties of the graphene and CNTs will also be summarised with their characterisation techniques.

3.2 SYNTHESIS OF GRAPHENE

Graphene has been synthesised in various ways depending on the research application. We summarise in Table 3.1 the different synthesis routes to produce single- or few-layer graphene and comment on their advantages and disadvantages.

3.2.1 MECHANICAL EXFOLIATION

Graphene was first discovered through a simple mechanical exfoliation (repeated peeling) by scotch tape technique from highly oriented pyrolytic graphite (HOPG). The advantages of mechanical exfoliation method are that graphene films realised by this method are monocrystalline, and no complicated experimental setup is needed. This method is the leading process for producing single-layer graphene flakes on preferred substrates. An external force of ~300 nN μm^{-2} is required to separate out one mono-atomic layer of graphene from graphite using the mechanical exfoliation method [14]. Different agents including scotch tape, ultrasonication, electric field, etc. were used for the peeling or exfoliation of graphene [15]. Although the quality of the graphene obtained by this method is very high with almost no defects, it is very difficult to obtain larger amounts of graphene by this exfoliation method.

3.2.2 CHEMICAL VAPOR DEPOSITION: SYNTHESIS AND TRANSFER

Chemical vapor deposition (CVD) is a useful procedure in which gas-phase molecules are decomposed and produce reactive species, which leads to the formation of film

TABLE 3.1
A Comparative Study of the Advantages and Disadvantages of Various Graphene Synthesis Methods

Synthesis Methods	Advantages	Disadvantages
Mechanical exfoliation	High-quality graphene; cheap equipment; easy operation	Low yield; small-sized graphene; time consuming
Chemically derived	Mass synthesis; graphene composites	Impurities; defects; hazardous reducing agents
Epitaxial growth on SiC	Mass synthesis; electronic devices	Defects; hard to obtain monolayer graphene; energy consuming
Chemical vapor deposition	High-quality graphene; high yield; large-sized graphene	Energy consuming; expensive
Arc deposition	Doped graphene	Expensive; low yield

or particle growth. CVD methods have been used for the deposition of a wide range of materials including conducting, semiconducting and insulating materials. High-quality graphene synthesis by the CVD procedure is typically done on various transition-metal substrates such as Cu, Ni, Pt, Ru, Ir, etc. [16]. Large-area growth of graphene was mainly practised on copper (Cu) and nickel (Ni) foil; however, the first attempt was made in 2008 on Ni substrate at 900°C [17].

Various types of CVD processes including thermal, plasma-enhanced CVD (PECVD), cold wall, hot wall, etc. are reported subject to the substrate quality, precursors, width, and structure required. The first CVD growth of graphene with a uniform as well as large area on polycrystalline Cu foils was demonstrated by using catalytic thermal decomposition of methane [18]. In this process, the predominantly single-layer graphene, with a minor percentage of less than 5% of the area having few layers, was obtained. Due to the few limiting properties of Cu including oxide formation, roughening and sublimation, researchers tried Ni by substituting Cu. For the growth process of graphene on Ni, first Ni foil was annealed by hydrogen followed by placing in a CH_4–Ar–H_2 environment for 20 minutes at a temperature of 1000°C [19]. It was found that the cooling rate controlled the thickness of the graphene layer. A faster cooling rate promoted the formation of a thicker layer, while the slower cooling rate avoided carbon segregation at the Ni surface [19]. The limiting factors of this process are that it is a time-consuming process and several wrinkles or folds interfere in the formation of homogeneous graphene.

3.2.2.1 Plasma-Enhanced Chemical Vapor Deposition

PECVD has emerged as an important method for producing graphene at a low temperature [20]. The possibility of energetic electrons developed by the plasma increasing the ionisation, excitation and dissociation of hydrocarbon precursors at a relatively low temperature can be realized [21]. Single- and few-layer graphene by PECVD on different substrates can be produced using 900-watt radio frequency (RF) power, 10 sccm total gas flow and an inside chamber pressure of ~12 Pa with 5%–100% CH_4 in H_2 gas mixture at 400°C–600°C substrate temperature [21]. The copper surface needs to be cleaned by a hydrogen plasma treatment prior to starting the graphene growth by this method [22].

3.2.2.2 Transfer of Graphene Sheet

There are several advantages of transferring graphene to another substrate. The graphene transfer technique solves the problem of separation of the graphene layer from the Cu/Ni substrate and shields the integrity of the graphene after separation [23]. As metal substrates can be effortlessly removed by etching solutions or peeled off after special pre-treatment, the key factor that regulates the quality of the transferred graphene is the support layer. The demand for large-scale, low-cost, efficient transfer of high-quality CVD graphene to arbitrary substrates has led to the growth of various optimised upgrades [24].

Graphene transfer was first done by using thermal release tape. In this method, low-residue graphene can be generated, but it is highly questionable if it can be produced in continuous coverage. The graphene transferred by the PMMA method has drawn tremendous attention. Poly(methyl methacrylate) (PMMA) has been used to transfer graphene to a substrate [25]. This method results in continuous graphene with minor residue because of the PMMA. Etching of the metal catalyst by a mild acidic solution is the wise choice to remove metal. It is considered that copper as well as nickel can be completely etched by 1 M solution of ferric chloride ($FeCl_3$) [26]. Substrates such as SiO_2, polydimethylsiloxane (PDMS), poly(ethylene terephthalate) (PET), etc. were used for the transfer of the graphene sheet.

3.2.3 Arc Discharge Method

Direct current arc discharge of graphite under a high pressure of hydrogen produces graphene flakes. This method typically produces graphene containing two to four layers in the inner wall region of the arc chamber [27]. The discharge current used for this method is in the range of 100–150 A, with a maximum open-circuit voltage of 60 V. Two electrodes were both 8 mm diameter pure graphite rods, and the current was kept at 120 A. The arc was kept by constantly transforming the cathode from the anode keeping a constant distance of 2 mm. N-doped graphene (size of 100–200 nm) with two to six layers can be synthesised by this method [28].

3.2.4 Epitaxial Growth

Epitaxial growth of single-layer graphene is a good technique [29,30]. It is the best alternative to mechanical exfoliation for the epitaxial growth of graphene on a hexagonal substrate. The first report was of ultrathin epitaxial graphene matured on single-crystal silicon carbide by vacuum graphitisation [31]. The as-grown sample of graphene produced by this method showed superior quality and was of larger scale than samples developed through conventional thermal desorption. In this method, graphene thickness can be controlled by changing the annealing temperature.

3.2.5 Chemical Synthesis

Chemical synthesis of reduced graphene oxide via reduction of graphite oxide (GO) is one of the most cost-effective approaches for large-scale application. The chemical methods for the synthesis of graphene involve two steps: exfoliation of graphite by oxidative treatment and reduction of the resulting GO using a suitable reducing agent. The synthesis procedure of GO by strong oxidation of graphite via chemical routes is the central need for chemical reduction. GO is highly hydrophilic and forms stable aqueous colloids, which can be converted through simple and cheap solution processes. The GO and reduced GO (rGO) are still interesting topics of research concerning the production of graphene, especially in regard to mass applications [32].

3.2.5.1 Synthesis of Graphite Oxide

The synthesis of GO is very interesting and historical. The famous methods used for the synthesis of GO are the Brodie method [33], the Staudenmaier method [34], the Hofmann method [35] and the Hummers method [36] and all the modified or improved methods. Briefly, graphite powder first chemically reacted with acids (HCl, H_2SO_4 and HNO_3, etc.) followed by the intercalation of alkali metals (alkali metal compounds $KClO_3$, $KMnO_4$, $NaNO_3$ etc.) into the graphitic layers which further helps in the exfoliation and breaking of graphitic layers into small pieces. Among them, modified Hummers method is the most used method followed by researchers, proved by the number of citations. The paper had 2466 citations and 84,011 article views (latest by September, 2019) according to Google scholar citations [36].

3.2.5.2 Reduction of Graphite Oxide

Chemical reduction of GO is one of the brilliant processes used to produce rGO and graphene in mass quantities. The reduction of graphene oxide is generally achieved using reducing agents such as hydrazine and its derivatives, $NaBH_4$, hydroquinone, sulfur compounds, metal/acid, etc. or by hydrothermal and solvothermal protocols [32,37–46]. The majority of reducing agents are either toxic or explosive and are difficult to handle for large-scale production. Moreover, most of the available methods are time consuming (6–48 hours) at room temperature or they require high temperatures [46]. The reduction of GO by metal/acid, including Fe/HCl, Al/HCl, Zn/H_2SO_4, Zn/HCl, etc. has drawn tremendous attention due to the fast reduction and high degree of reduction obtained [10]. Chemical reduction of GO using Zn metal and mineral acid at room temperature is rapid (2 hours) and does not involve the use of any other toxic reagents [10].

3.2.6 ELECTROCHEMICAL ROUTE

Synthesis of reduced graphene oxide/graphene by the electrochemical method can be achieved by two different routes. One can first obtain GO from graphite by several methods stated earlier followed by electrochemical reduction of GO [47–50]. Another route for the synthesis of graphene involves electrochemical exfoliation from graphite and is considered to be a fast and environmentally friendly method of production of high-quality graphene [51,52]. An electrochemical reduction of colloidal GO solution was carried out using potential scanning under applied DC bias [50,53,54] or at constant working potential [49,55]. Electrochemical exfoliation can produce high-quality graphene from HOPG in an acidic medium [52] or through graphite rods in aqueous $(NH_4)_2SO_4$ solution [51] using a two-electrode system under constant current. An aqueous solution of graphene or electrochemically reduced graphene oxide produced using the electrochemical route is highly pure, and it is a more rapid process than any other method.

3.2.7 BIOMASS-/WASTE-DERIVED GRAPHENE

The waste material originating from biomass is esteemed not only for possible sustainable energy resources, but also for possible use as carbon sources like graphene. Recently, biomass wastes such as coconut shells were used to produce porous graphene nanosheets [56]. In this method, $ZnCl_2$ was mixed with $FeCl_3$ solution to produce an activating agent. Ding et al. recently reported both cathode and anode materials of a hybrid sodium ion capacitor derived from peanut shells. He et al. developed a supercapacitor with mesoporous carbon derived from peanut biomass using $ZnCl_2$ activation–assisted microwave heating.

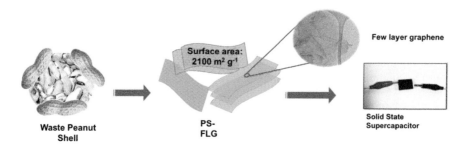

FIGURE 3.1 Few-layer graphene sheets derived from waste peanut shells.

Recently, a rather common method of mechanical exfoliation was reformulated, introducing the utilisation of no-use peanut shells to obtain value-added few-layer graphene (FLG)–like sheets (Figure 3.1) with an extensive surface area for the development of high-energy and high-power supercapacitors [57]. This procedure is highly competent, environmentally friendly, cost-effective, scalable and most importantly it does not involve any chemical graphitising reagent/catalyst.

3.2.8 OTHER SYNTHESIS METHODS

Various other fabrication methods are also available for graphene synthesis.

The unzipping of SWCNTs or MWCNTs with oxidising mediators was used for the synthesis of graphene. The electrochemical unzipping of MWCNTs into large-aspect-ratio N-doped graphene nanoribbons can be achieved in a single-step electrochemical process at room temperature in formamide medium (used as a solvent and a source of nitrogen) [58].

Recent reports show that laser irradiation or plasma etching is a way to produce graphene [59]. Spontaneous exfoliation of graphite through intercalation compound methods has also been used to obtain graphene [60].

Exfoliation of graphite into graphene or functionalised graphene was made possible by using pyrene and porphyrin in organic medium based on the $\pi-\pi$ interaction among graphene and porphyrins/pyrene molecules [60–63].

3.3 SYNTHESIS OF CARBON NANOTUBES

3.3.1 ARC DISCHARGE

CNTs were first synthesised using the arc-discharge method by Iijima in 1991 [64] using the experimental setup illustrated in Figure 3.2a and conditions similar to those for the synthesis of fullerenes. Two graphite electrodes are placed close to each other (1 mm) in inert atmosphere at a pressure of 500 torr. A voltage of 20–25 V with a DC current of 50–120 A is desired to generate arc between the electrodes [65]. In this condition, a very high temperature in the chamber evaporates carbon from the electrodes. The deposition of CNT occurs on the cathode by re-condensation of the arc-evaporated material. The SWCNTs are produced by this method on the electrode with metals such as Ni, Fe, Co, Gd and Y. Although both SWCNTs and MWCNTs can

Synthesis and Characterisation of Carbonaceous Materials

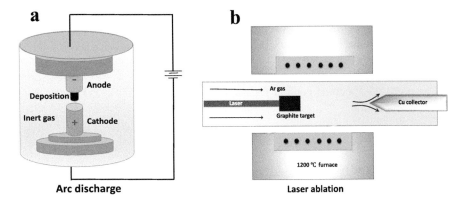

FIGURE 3.2 (a) Arc discharge setup. (b) Laser ablation system.

be produced by this method with moderate cost, the disadvantages of this technique are carbon impurities and encapsulated nanoparticles that are usually produced in addition to CNTs.

Several methods have been reported for the large-scale production of CNTs by arc discharge technique [66]. One can play with the vapor pressure of the catalyst metal to achieve high yield of the CNT [67,68]. Large-scale synthesis of CNTs was recently demonstrated by using low-pressure flowing air as buffer gas during DC arc discharge, where iron was used as a catalyst and sulfur was used as a promoter [66].

3.3.2 Laser Ablation

CNTs are synthesised using the laser ablation method by using a target doped with metals such as Ni, Co and Pt. In this method, high temperature is generated by a laser beam to gasify the carbon target [69]. The graphite target is placed in a quartz tube surrounded by a furnace heated at 800°C–1500°C and 500 torr of argon gas is passed through the tube to carry the soot formed to a water-cooled Cu collector (Figure 3.2b). The disadvantage of this method is that produced CNTs are in carbonaceous soot, and impurities such as metal catalysts and amorphous carbon exist as by-products [70]. A modified method was reported to grow SWCNTs with deposited nanoparticles catalyst on SiO_2/Si substrates using pulsed laser deposition (PLD) and exposing them to the carbon vapor produced by the KrF-laser ablation of a natural graphite target [65].

3.3.3 Chemical Vapor Deposition Method

CNTs are synthesised from hydrocarbons, such as methane, which was first observed being adsorbed on a catalytic metal surface followed by the decomposition of carbon atoms under Argon atmospheres. In the CVD method, carbon atoms are first diffused into catalyst substrate, and then as soon as a supersaturated state is achieved, the precipitated carbon atoms serve as seed points for nucleation and growth of CNTs [71,72]. CNT growth at low temperatures such as 600°C–900°C generally yields

MWCNTs, whereas comparatively higher temperatures such as 900°C–1200°C helps growth of SWCNTs [73]. The most popular metals including iron, nickel, cobalt and molybdenum are used to produce CNTs. In case of CVD grown CNT, Cu should not be used as metal catalyst because it has low catalytic activity due to its nearly zero carbon solubility. Recently metal-free catalysts such as diamond, Al_2O_3, SiO_2, ZrO_2 and graphite have become popular for the production of CNTs due to large-scale at low-cost production.

3.4 CHARACTERISATIONS

Graphene and CNTs synthesised from different procedures are characterised through various analytical techniques. The structures of different carbon allotropes are classified, described and illustrated by different methods such as microscopic technique (transmission electron microscopy [TEM], scanning electron microscopy [SEM] and atomic force microscopy [AFM]), X-ray diffraction, Raman spectra, ultraviolet visible (UV-vis), Fourier transform infrared (FTIR) and X-ray photoelectron spectroscopy as summarised here.

3.4.1 Microscopic Technique

AFM is a powerful technique for investigating the structure of carbonaceous material due to its high spatial resolution and the various modes that allow for probing of different physical properties. AFM traditionally operates in two modes depending on the way the tip moves, such as contact and tapping modes. If the sample surface is damaged without difficulty by dragging the tip across it, then tapping mode is preferred as it avoids friction, adhesion and electrostatic forces by oscillating the tip, which then moves on and off the surface to provide a high-resolution image. However, in contact mode (static mode), the tip is 'drawn' across the sample in contact with the surface, which is mapped by recording the deflection of the cantilever directly. Graphene is a two-dimensional single-layer hexagonal lattice of tightly bonded carbon atoms in a sp^2 hybridised arrangement and has a single-atom thickness of around 3.45 Å [74]. AFM measurement of graphene provides the topographical information, which directly helps to obtain the thickness of the graphene layer. However, literature reports of AFM offer a wide range of measured values for single-layer graphene thickness between 0.4 and 1.7 nm [75]. This inconsistency has been ascribed to tip-surface interactions, image feedback settings and surface chemistry.

SEM and TEM are used to validate the surface morphology of carbonaceous materials like graphene and CNTs. TEM analysis also proves the flake-like shapes of reduced GO. Single-layer transparent graphene sheets can be perceived by TEM analysis. Hernandez et al. have shown that the thickness of graphene can be determined accurately by TEM analysis by observing a large number of TEM images to generate a series of thickness statistics [76]. The selected area diffraction pattern (SAED) confirms the disordered and crystalline nature of the rGO/graphene nanosheets. SAED is a useful technique for accurately determining the number of graphene layers by nanoarea electron diffraction patterns through changing incidence angles between the electron beam and the graphene sheet [65,76].

3.4.2 RAMAN ANALYSIS

Raman spectroscopic investigations are an important tool to investigate the structure of carbon like graphene and CNTs through a non-destructive method. Raman analysis of carbon allotropes showed their peaks of D, G, 2D and S3 around 1350, 1580, 2700 and 2950 cm^{-1}, respectively [77]. Generally, the quality of graphene, specifically for determining defects and ordered/disordered structures of graphene and number of layers that exist in the sample can be identified by Raman spectroscopy [78]. The G band of graphene is ascribed to the $E_{2}g$ phonon of C sp^2 atoms, whereas the D band corresponds to the breathing mode of κ-point phonons of A_1g symmetry [74,79,80]. The second-order (two-photon) spectrum of graphene, ascribed as a 2D band arises from the overtone (second harmonic) of the D band [74]. The monolayer graphene structure can be verified from the intensity ratio of G band and 2D band, I_G/I_{2D} [81]. The intensity ratio of D to G band, I_D/I_G is usually used to characterise the defects and disorders in both the graphene and the CNT samples [81,82]. The crystalline size (L_a) of a graphitic sample is inversely proportional to the D to G band intensity (I_D/I_G) [83].

The in-plane crystallite sizes (L_a) of the undoped and doped graphitic samples can be calculated by the following formula [80,83]:

$$L_a(nm) = (2.4 \times 10^{-10})\lambda^4 (I_D/I_G)^{-1}$$

Here λ is the wavelength used for Raman measurements, and I_D, I_G are the intensities of D and G band, respectively.

3.4.3 X-RAY DIFFRACTION

The crystalline nature of the graphite, graphene oxide graphene and CNTs as well as d-spacing can be characterised by XRD measurement [84–86]. The XRD pattern of the graphite shows a well-defined (002) peak at 26.4° that corresponds to the d-spacing of 0.335 nm, whereas MWCNT shows a peak at around 25.3° [83]. The (002) peak for GO shows at 10°, which corresponds to the d-spacing of 0.88 nm [87,88]. A reduced GO has its peak at around 23.8° with d-spacing of 0.34 nm, indicating more interlayer spacing than graphite but less interlayer spacing than GO [83].

3.4.4 X-RAY PHOTOELECTRON SPECTROSCOPY

X-ray photoelectron spectroscopy (XPS) spectra are very useful in identifying the atomic percentages of C and any other heteroatoms if present in the carbonaceous architecture. The atomic ratio of C/O, obtained from C1s and O1s peak areas, is often very useful for the GO and reduced GO to characterise the samples [74]. The C1s XPS spectrum of GO confirms the degree of oxidation of C atoms in the form of different functional groups. The different functional group of C with O atoms, such as the C atom in the C–O bond at 286.3 eV, the non-oxygenated ring C=C at 284.3 eV, the carbonyl C (C=O) at 288.1 eV, and the carboxylate carbon (HO–C=O) at 289.4 eV can be isolated from C1s XPS spectra [89]. All of these peaks except the peak at 284.3 eV for C=C are either reduced in intensity or almost vanish after reduction of GO to rGO.

3.5 CONCLUSION

Carbonaceous materials like graphene and carbon nanotubes are the important part of any energy device and technology. The synthesis of these two materials sometimes plays an important role in making efficient devices. Various methods are demonstrated in this chapter to provide a brief discussion of their synthesis route and their advantages and disadvantages. Characterisations of the synthesis materials play an important role in understanding the structures of the materials and their properties. Several characterisation techniques such as microscopic characterisations, Raman, XPS, XRD, etc. are used to characterise the graphene, and CNTs are also demonstrated in this chapter. High-resolution characterisation practices help in determining the actual structures that are useful to understand the physical and chemical properties of carbonaceous materials.

REFERENCES

1. W. Choi, I. Lahiri, R. Seelaboyina, Y.S. Kang, Synthesis of graphene and its applications: A review, *Crit. Rev. Solid State Mater. Sci.* 35, 2010, 52–71. doi:10.1080/10408430903505036
2. S. Chen, W. Cai, R.D. Piner, J.W. Suk, Y. Wu, Y. Ren, J. Kang, R.S. Ruoff, Synthesis and characterization of large-area graphene and graphite films on commercial Cu-Ni alloy foils, *Nano Lett.* 11, 2011, 3519–3525. doi:10.1021/nl201699j
3. W. Zhang, Y. Zhang, Y. Tian, Z. Yang, Q. Xiao, X. Guo, L. Jing et al. Insight into the capacitive properties of reduced graphene oxide, *ACS Appl. Mater. Interfaces.* 6, 2014, 2248–2254. doi:10.1021/am4057562
4. B.C.-K. Tee, C. Wang, R. Allen, Z. Bao, An electrically and mechanically self-healing composite with pressure- and flexion-sensitive properties for electronic skin applications, *Nat. Nanotechnol.* 7, 2012, 825–832. doi:10.1038/nnano.2012.192
5. P. Kumar, A. Kumar Singh, S. Hussain, K. Nam Hui, K. San Hui, J. Eom, J. Jung, J. Singh, Graphene: Synthesis, properties and application in transparent electronic devices, *Rev. Adv. Sci. Eng.* 2, 2013, 238–258. doi:10.1166/rase.2013.1043
6. S. Stankovich, D.A. Dikin, G.H.B. Dommett, K.M. Kohlhaas, E.J. Zimney, E.A. Stach, R.D. Piner, S.B.T. Nguyen, R.S. Ruoff, Graphene-based composite materials, *Nature.* 442, 2006, 282–286. doi:10.1038/nature04969
7. C.K. Chua, M. Pumera, Covalent chemistry on graphene, *Chem. Soc. Rev.* 42, 2013, 3222–3233. doi:10.1039/c2cs35474h
8. K.S. Novoselov, V.I. Fal, L. Colombo, P.R. Gellert, M.G. Schwab, K. Kim, V.I.F. Ko et al. A roadmap for graphene, *Nature.* 490, 2013, 192–200. doi:10.1038/nature11458
9. D.R. Dreyer, S. Park, C.W. Bielawski, R.S. Ruoff, The chemistry of graphene oxide, *Chem. Soc. Rev.* 39, 2010, 228–240. doi:10.1039/b920539j
10. R.S. Dey, S. Hajra, R.K. Sahu, C.R. Raj, M.K. Panigrahi, A rapid room temperature chemical route for the synthesis of graphene: Metal-mediated reduction of graphene oxide, *Chem. Commun.* 48, 2012, 1787. doi:10.1039/c2cc16031e
11. X. Li, H. Wang, J.T. Robinson, H. Sanchez, G. Diankov, H. Dai, Simultaneous nitrogen doping and reduction of graphene oxide, *J. Am. Chem. Soc.* 131, 2009, 15939–15944. doi:10.1021/ja907098f

12. S. Banerjee, S.S. Wong, Synthesis and characterization of carbon nanotube–nanocrystal heterostructures, *Nano Lett.* 2, 2002, 195–200. doi:10.1021/nl015651n
13. G.A. Rivas, M.D. Rubianes, M.C. Rodríguez, N.F. Ferreyra, G.L. Luque, M.L. Pedano, S.A. Miscoria, C. Parrado, Carbon nanotubes for electrochemical biosensing, *Talanta.* 74, 2007, 291–307. doi:10.1016/j.talanta.2007.10.013
14. Y. Zhang, J.P. Small, W.V. Pontius, P. Kim, Fabrication and electric-field-dependent transport measurements of mesoscopic graphite devices, *Appl. Phys. Lett.* 86, 2005, 1–3. doi:10.1063/1.1862334
15. M.S.A. Bhuyan, M.N. Uddin, M.M. Islam, F.A. Bipasha, S.S. Hossain, Synthesis of graphene, *Int. Nano Lett.* 6, 2016, 65–83. doi:10.1007/s40089-015-0176-1
16. A. Reina, X. Jia, J. Ho, D. Nezich, H. Son, V. Bulovic, M.S. Dresselhaus, J. Kong, Large area, few-layer graphene films on arbitrary substrates by chemical vapor deposition, *Nano Lett.* 9, 2009, 30–35. doi:10.1021/nl801827v
17. A.N. Obraztsov, E.A. Obraztsova, A.V. Tyurnina, A.A. Zolotukhin, Chemical vapor deposition of thin graphite films of nanometer thickness, *Carbon N. Y.* 45, 2007, 2017–2021. doi:10.1016/j.carbon.2007.05.028
18. X. Li, W. Cai, J. An, S. Kim, J. Nah, D. Yang, R. Piner et al. Large area synthesis of high quality and uniform graphene films on copper foils, *Science.* 324, 2009, 1312–1314. doi:10.1126/science.1171245
19. Q. Yu, J. Lian, S. Siriponglert, H. Li, Y.P. Chen, S.S. Pei, Graphene segregated on Ni surfaces and transferred to insulators, *Appl. Phys. Lett.* 93, 2008, 1–4. doi:10.1063/1.2982585
20. M. Li, D. Liu, D. Wei, X. Song, D. Wei, A.T.S. Wee, Controllable synthesis of graphene by plasma-enhanced chemical vapor deposition and its related applications, *Adv. Sci.* 3, 2016, 1–23. doi:10.1002/advs.201600003
21. Z. Bo, Y. Yang, J. Chen, K. Yu, J. Yan, K. Cen, Plasma-enhanced chemical vapor deposition synthesis of vertically oriented graphene nanosheets, *Nanoscale.* 5, 2013, 5180–5204. doi:10.1039/c3nr33449j
22. N. Woehrl, O. Ochedowski, S. Gottlieb, K. Shibasaki, S. Schulz, Plasma-enhanced chemical vapor deposition of graphene on copper substrates, *AIP Adv.* 4, 2014, 1–9. doi:10.1063/1.4873157
23. L. Imperiali, C. Clasen, J. Fransaer, C.W. Macosko, J. Vermant, A simple route towards graphene oxide frameworks, *Mater. Horiz.* 1, 2014, 139–145. doi:10.1039/C3MH00047H
24. A. Boscá, J. Pedrós, J. Martínez, T. Palacios, F. Calle, Automatic graphene transfer system for improved material quality and efficiency, *Sci. Rep.* 6, 2016, 1–8. doi:10.1038/srep21676
25. X. Li, Y. Zhu, W. Cai, M. Borysiak, B. Han, D. Chen, R.D. Piner, L. Colombo, R.S. Ruoff, Transfer of large-area graphene films for high-performance transparent conductive electrodes, *Nano Lett.* 9, 2009, 4359–4363. doi:10.1021/nl902623y
26. X. Liang, B.A. Sperling, I. Calizo, G. Cheng, C.A. Hacker, Q. Zhang, Y. Obeng et al. Toward clean and crackless transfer of graphene, *ACS Nano.* 5, 2011, 9144–9153. doi:10.1021/nn203377t
27. K.S. Subrahmanyam, L.S. Panchakarla, A. Govindaraj, C.N.R. Rao, Simple method of preparing graphene flakes by an arc-discharge method, *J. Phys. Chem. C.* 113, 2009, 4257–4259. doi:10.1021/jp900791y
28. N. Li, Z. Wang, K. Zhao, Z. Shi, Z. Gu, S. Xu, Large scale synthesis of N-doped multi-layered graphene sheets by simple arc-discharge method, *Carbon N. Y.* 48, 2009, 255–259.
29. H. Tetlow, J. Posthuma de Boer, I.J. Ford, D.D. Vvedensky, J. Coraux, L. Kantorovich, Growth of epitaxial graphene: Theory and experiment, *Phys. Rep.* 542, 2014, 195–295. doi:10.1016/j.physrep.2014.03.003

30. J.L. Qi, K. Nagashio, T. Nishimura, A. Toriumi, Crystal orientation relation and macroscopic surface roughness in hetero-epitaxially grown graphene on Cu/mica, *Nanotechnology.* 25, 2014, 185602. doi:10.1088/0957-4484/25/18/185602
31. C. Berger, Z. Song, X. Li, X. Wu, N. Brown, C. Naud, D. Mayou et al. Electronic confinement and coherence in patterned epitaxial graphene, *Science* 312, 2006, 1191–1196. doi:10.1126/science.1125925
32. R.K. Singh, R. Kumar, D.P. Singh, Graphene oxide: Strategies for synthesis, reduction and frontier applications, *RSC Adv.* 6, 2016, 64993–65011. doi:10.1039/c6ra07626b
33. B.C. Brodie, On the atomic weight of graphite, *Philos. Trans. R. Soc. London.* 149, 1859, 249–259. doi:10.1098/rstl.1859.0013
34. L. Staudenmaier, Verfahren zur Darstellung der Graphitsaure, *Berichte Der Dtsch. Chem. Gesellschaft.* 31, 1898, 1481–1487. doi:10.1002/cber.18980310237
35. U. Hofmann, E. König, Studies on graphite oxide, *J. Inorg. Gen. Chem.* 234, 1937, 311–336. doi:10.1002/zaac.19372340405
36. W.S. Hummers, R.E. Offeman, Preparation of graphitic oxide, *J. Am. Chem. Soc.* 80, 1958, 1339. doi:10.1021/ja01539a017
37. J. Ning, J. Wang, X. Li, T. Qiu, B. Luo, L. Hao, M. Liang, B. Wang, L. Zhi, A fast room-temperature strategy for direct reduction of graphene oxide films towards flexible transparent conductive films, *J. Mater. Chem. A.* 2, 2014, 10969–10973. doi:10.1039/c4ta00527a
38. T. Lu, L. Pan, C. Nie, Z. Zhao, Z. Sun, A green and fast way for reduction of graphene oxide in acidic aqueous solution via microwave assistance, *Phys. Status Solidi Appl. Mater. Sci.* 208, 2011, 2325–2327. doi:10.1002/pssa.201084166
39. S. Dubin, S. Gilje, K. Wang, V.C. Tung, K. Cha, A.S. Hall, J. Farrar, R. Varshneya, Y. Yang, R.B. Kaner, A one-step, solvothermal reduction method for producing reduced graphene oxide dispersions in organic solvents, *ACS Nano.* 4, 2010, 3845–3852. doi:10.1021/nn100511a
40. V.P. Novikov, S.A. Kirik, Low-temperature synthesis of graphene, *Tech. Phys. Lett.* 37, 2011, 565–567. doi:10.1134/S1063785011060265
41. T. Kuila, S. Bose, P. Khanra, A.K. Mishra, N.H. Kim, J.H. Lee, A green approach for the reduction of graphene oxide by wild carrot root, *Carbon N. Y.* 50, 2012, 914–921. doi:10.1016/j.carbon.2011.09.053
42. J. Gao, F. Liu, Y. Liu, N. Ma, Z. Wang, X. Zhang, Environment-friendly method to produce graphene that employs vitamin C and amino acid, *Chem. Mater.* 22, 2010, 2213–2218. doi:10.1021/cm902635j
43. V.H. Pham, T.V. Cuong, T.D. Nguyen-Phan, H.D. Pham, E.J. Kim, S.H. Hur, E.W. Shin, S. Kim, J.S. Chung, One-step synthesis of superior dispersion of chemically converted graphene in organic solvents, *Chem. Commun.* 46, 2010, 4375–4377. doi:10.1039/c0cc00363h
44. I.K. Moon, J. Lee, H. Lee, Highly qualified reduced graphene oxides: The best chemical reduction, *Chem. Commun.* 47, 2011, 9681–9683. doi:10.1039/c1cc13312h
45. X. Yan, X. Cui, L.S. Li, Synthesis of large, stable colloidal graphene quantum dots with tunable size, *J. Am. Chem. Soc.* 132, 2010, 5944–5945. doi:10.1021/ja1009376
46. J. Zhang, H. Yang, G. Shen, P. Cheng, J. Zhang, S. Guo, Reduction of graphene oxide vial-ascorbic acid, *Chem. Commun.* 46, 2010, 1112–1114. doi:10.1039/b917705a
47. B. Devadas, M. Rajkumar, Electrochemically reduced graphene oxide/neodymium hexacyanoferrate modified electrodes for the electrochemical detection of paracetomol, *Int. J. Electrochem. Sci.* 7, 2012, 3339–3349.
48. M. Zhou, Y. Wang, Y. Zhai, J. Zhai, W. Ren, F. Wang, S. Dong, Controlled synthesis of large-area and patterned electrochemically reduced graphene oxide films, *Chem. A Eur. J.* 15, 2009, 6116–6120. doi:10.1002/chem.200900596

49. J. Ping, Y. Wang, Y. Ying, J. Wu, Application of electrochemically reduced graphene oxide on screen-printed ion-selective electrode, *Anal. Chem.* 84, 2012, 3473–3479. doi:10.1021/ac203480z
50. Y. Shao, J. Wang, M. Engelhard, C. Wang, Y. Lin, Facile and controllable electrochemical reduction of graphene oxide and its applications, *J. Mater. Chem.* 20, 2010, 743. doi:10.1039/b917975e
51. K. Parvez, R.A. Rincón, N.E. Weber, K.C. Cha, S.S. Venkataraman, One-step electrochemical synthesis of nitrogen and sulfur co-doped, high-quality graphene oxide, *Chem. Commun.* 52, 2016, 5714–5717. doi:10.1039/c6cc01250g
52. C.Y. Su, A.Y. Lu, Y. Xu, F.R. Chen, A.N. Khlobystov, L.J. Li, High-quality thin graphene films from fast electrochemical exfoliation, *ACS Nano.* 5, 2011, 2332–2339. doi:10.1021/nn200025p
53. G.K. Ramesha, N.S. Sampath, Electrochemical reduction of oriented Graphene oxide films: An in situ Raman spectroelectrochemical study, *J. Phys. Chem. C.* 113, 2009, 7985–7989. doi:10.1021/jp811377n
54. Z. Wang, X. Zhou, J. Zhang, F. Boey, H. Zhang, Direct electrochemical reduction of single-layer graphene oxide and subsequent functionalization with glucose oxidase, *J. Phys. Chem. C.* 113, 2009, 14071–14075. doi:10.1021/jp906348x
55. X.Y. Peng, X.X. Liu, D. Diamond, K.T. Lau, Synthesis of electrochemically-reduced graphene oxide film with controllable size and thickness and its use in supercapacitor, *Carbon N. Y.* 49, 2011, 3488–3496. doi:10.1016/j.carbon.2011.04.047
56. L. Sun, C. Tian, M. Li, X. Meng, L. Wang, R. Wang, J. Yin, H. Fu, From coconut shell to porous graphene-like nanosheets for high-power supercapacitors, *J. Mater. Chem. A.* 1, 2013, 6462. doi:10.1039/c3ta10897j
57. T. Purkait, G. Singh, M. Singh, D. Kumar, R.S. Dey, Large area few-layer graphene with scalable preparation from waste biomass for high-performance supercapacitor, *Sci. Rep.* 7, 2017, 15239. doi:10.1038/s41598-017-15463-w
58. M.J. Jaison, T.N. Narayanan, T. Prem Kumar, V.K. Pillai, A single-step room-temperature electrochemical synthesis of nitrogen-doped graphene nanoribbons from carbon nanotubes, *J. Mater. Chem. A.* 3, 2015, 18222–18228. doi:10.1039/c5ta03869c
59. S. Lee, M.F. Toney, W. Ko, J.C. Randel, H.J. Jung, K. Munakata, J. Lu et al. Laser-synthesized epitaxial graphene, *ACS Nano.* 4, 2010, 7524–7530. doi:10.1021/nn101796e
60. X. An, T. Simmons, R. Shah, C. Wolfe, K.M. Lewis, M. Washington, S.K. Nayak, S. Talapatra, S. Kar, Stable aqueous dispersions of noncovalently functionalized graphene from graphite and their multifunctional high-performance applications, *Nano Lett.* 10, 2010, 4295–4301. doi:10.1021/nl903557p
61. J. Geng, B.S. Kong, S.B. Yang, H.T. Jung, Preparation of graphene relying on porphyrin exfoliation of graphite, *Chem. Commun.* 46, 2010, 5091–5093. doi:10.1039/c001609h
62. J. Malig, A.W.I. Stephenson, P. Wagner, G.G. Wallace, D.L. Officer, D.M. Guldi, Direct exfoliation of graphite with a porphyrin – Creating functionalizable nanographene hybrids, *Chem. Commun.* 48, 2012, 8745–8747. doi:10.1039/c2cc32888g
63. M. Cardinali, L. Valentini, J.M. Kenny, Anisotropic electrical transport properties of graphene nanoplatelets/pyrene composites by electric field assisted thermal annealing, *J. Phys. Chem. C.* 33, 2011 16652–16656. doi: 10.1021/jp205772x.
64. S. Iijima, Helical microtubules of graphitic carbon, *Nature.* 354, 1991, 56–58. doi:10.1038/354056a0
65. W.W. Liu, S.P. Chai, A.R. Mohamed, U. Hashim, Synthesis and characterization of graphene and carbon nanotubes: A review on the past and recent developments, *J. Ind. Eng. Chem.* 20, 2014, 1171–1185. doi:10.1016/j.jiec.2013.08.028
66. Y. Su, P. Zhou, J. Zhao, Z. Yang, Y. Zhang, Large-scale synthesis of few-walled carbon nanotubes by DC arc discharge in low-pressure flowing air, *Mater. Res. Bull.* 48, 2013, 3232–3235. doi:10.1016/j.materresbull.2013.04.092

67. Z. Shi, Y. Lian, X. Zhou, Z. Gu, Y. Zhang, S. Iijima, L. Zhou, K.T. Yue, S. Zhang, Mass-production of single-wall carbon nanotubes by arc discharge method, *Carbon N. Y.* 37, 1999, 1449–1453. doi:10.1016/S0008-6223(99)00007-X
68. C. Liu, H.M. Cheng, Carbon nanotubes: Controlled growth and application, *Mater. Today.* 16, 2013, 19–28. doi:10.1016/j.mattod.2013.01.019
69. J. Chrzanowska, J. Hoffman, A. Małolepszy, M. Mazurkiewicz, T.A. Kowalewicz, Z. Szymanski, L. Stobinski, Synthesis of carbon nanotubes by the laser ablation method: Effect of laser wavelength, *Phys. Status Solidi.* 252, 2015, 1860–1867. doi:10.1002/pssb.201451614
70. T.F. Kuo, C.C. Chi, I.N. Lin, Synthesis of carbon nanotubes by laser ablation of graphites at room temperature, *Jpn. J. Appl. Phys.* 40, 2001, 7147–7150.
71. L.C. Qin, CVD synthesis of carbon nanotubes, *J. Mater. Sci. Lett.* 16, 1997, 457–459. doi:10.1023/A:1018504108114
72. A. Magrez, J.W. Seo, R. Smajda, M. Mionić, L. Forró, Catalytic CVD synthesis of carbon nanotubes: Towards high yield and low temperature growth, *Materials (Basel).* 3, 2010, 4871–4891. doi:10.3390/ma3114871
73. S.A. Bakar, A.A. Aziz, P. Marwoto, S. Sakrani, Amorphous hydrogenated carbon films, *Carbon N. Y,* 77, 1987, 311–320. doi:10.1007/8611
74. J.I. Paredes, S. Villar-Rodil, P. Solís Fernández, A. Martínez-Alonso, J.M.D. Tascón, Atomic force and scanning tunneling microscopy imaging of graphene nanosheets derived from graphite oxide, *Langmuir.* 25, 2009, 5957–5968. doi:10.1021/la804216z
75. C.J. Shearer, A.D. Slattery, A.J. Stapleton, J.G. Shapter, C.T. Gibson, Accurate thickness measurement of graphene, *Nanotechnology.* 27, 2016, 125704. doi:10.1088/0957-4484/27/12/125704
76. V. Nicolosi, R. Goodhue, B. Holland, V. Scardaci, J. Hutchison, S. Krishnamurthy, A.C. Ferrari et al. High-yield production of graphene by liquid-phase exfoliation of graphite, *Nat. Nanotechnol.* 3, 2008, 563–568. doi:10.1038/nnano.2008.215
77. V.C. Tung, M.J. Allen, Y. Yang, R.B. Kaner, High-throughput solution processing of large-scale graphene, *Nat. Nanotechnol.* 4, 2009, 25–29. doi:10.1038/nnano.2008.329
78. I.K. Moon, J. Lee, R.S. Ruoff, H. Lee, Reduced graphene oxide by chemical graphitization, *Nat. Commun.* 1, 2010, 73–4. doi:10.1038/ncomms1067
79. C.E. Hamilton, J.R. Lomeda, Z. Sun, J.M. Tour, A.R. Barron, High-yield organic dispersions of unfunctionalized graphene, *Nano Lett.* 9, 2009, 3460–3462. doi:10.1021/nl9016623
80. L.S. Panchakarla, K.S. Subrahmanyam, S.K. Saha, A. Govindaraj, H.R. Krishnamurthy, U.V. Waghmare, C.N.R. Rao, Synthesis, structure, and properties of boron- and nitrogen-doped graphene, *Adv. Mater.* 21, 2009, 4726–4730. doi:10.1002/adma.200901285
81. G.P. Dai, J.M. Zhang, S. Deng, Synthesis and characterization of nitrogen-doped monolayer and multilayer graphene on TEM copper grids, *Chem. Phys. Lett.* 516, 2011, 212–215. doi:10.1016/j.cplett.2011.09.081
82. Q. Meng, H. Wu, Y. Meng, K. Xie, Z. Wei, Z. Guo, High-performance all-carbon yarn micro-supercapacitor for an integrated energy system, *Adv. Mater.* 26, 2014, 4100–4106. doi:10.1002/adma.201400399
83. L.G. Cançado, K. Takai, T. Enoki, M. Endo, Y.A. Kim, H. Mizusaki, A. Jorio, L.N. Coelho, R. Magalhães-Paniago, M.A. Pimenta, General equation for the determination of the crystallite size L_a of nanographite by Raman spectroscopy, *Appl. Phys. Lett.* 88, 2006, 1–4. doi:10.1063/1.2196057
84. K.H. Lee, S.W. Han, K.Y. Kwon, J.B. Park, Systematic analysis of palladium-graphene nanocomposites and their catalytic applications in Sonogashira reaction, *J. Colloid Interface Sci.* 403, 2013, 127–133. doi:10.1016/j.jcis.2013.04.006

85. W.R. Collins, W. Lewandowski, E. Schmois, J. Walish, T.M. Swager, Claisen rearrangement of graphite oxide: A route to covalently functionalized graphenes, *Angew. Chemie Int. Ed.* 50, 2011, 8848–8852. doi:10.1002/anie.201101371
86. M. Herrera-Alonso, A.A. Abdala, M.J. McAllister, I.A. Aksay, R.K. Prud'homme, Intercalation and stitching of graphite oxide with diaminoalkanes, *Langmuir.* 23, 2007, 10644–10649. doi:10.1021/la0633839
87. S. Stankovich, D.A. Dikin, O.C. Compton, G.H.B. Dommett, R.S. Ruoff, S.T. Nguyen, Systematic post-assembly modification of graphene oxide paper with primary alkylamines, *Chem. Mater.* 22, 2010, 4153–4157. doi:10.1021/cm100454g
88. Y. Yang, T. Liu, Fabrication and characterization of graphene oxide/zinc oxide nanorods hybrid, *Appl. Surf. Sci.* 257, 2011, 8950–8954. doi:10.1016/j.apsusc.2011.05.070
89. Y.E. Shin, Y.J. Sa, S. Park, J. Lee, K.H. Shin, S.H. Joo, H. Ko, An ice-templated, pH-tunable self-assembly route to hierarchically porous graphene nanoscroll networks, *Nanoscale.* 6, 2014, 9734–9741. doi:10.1039/c4nr01988a

4 Rechargeable Battery Technology

4.1 INTRODUCTION

A device that transduces chemical energy to electrical energy and vice versa is known as a battery. It comprises a cathode, an anode, and the electrolyte (Figure 4.1). In the case of a lithium (Li) battery, the source of lithium ions is an anode, whereas the cathode is the sink for the lithium ions. The role of the electrolyte in a battery is the separation of ionic transport and electronic transport. The cell electrochemical potential is determined by the difference between the chemical potential of the metal ion in the anode and cathode, $\Delta G = -nFE$ [1], where ΔG represents change in internal energy during ion and electron insertion, n is the number of electrons stored, F is the faraday constant, and E stands for the cell potential.

The lithium-ion battery (LIB) is a state-of-the-art rechargeable power source for most electrical and electronic devices. The lithium ion–intercalated graphite is considered as the first example of lithium storage in a carbon-based nanomaterial, although Schaffäutl contributed the first report on carbon-based intercalation compounds in 1841 [2,3]. The electrode in the LIB must be both a good ionic conductor as well as a good electronic conductor, because the Li-ion in LIB flows through the electrolyte, whereas the electron generated from the reaction $Li = Li^+ + e^-$ passes through the external circuit [4]. Therefore, it is clear that the electrode material must have good electronic conductivity to act as an electrode-active material. As the effective radius of the Li-ion is small in comparison with other monovalent ions apart from H^+, it is an ideal ionic carrier for the transportation of electronic charges into various matrixes. Metallic lithium is being used as an anode in LIBs and is found to be somewhat of a problem because of dendrite formation during its long cycling which may cause short circuit of the battery [1]. The growing demand for portable electronics has pushed the market for LIBs, resulting in one-quarter of the world's production of the precursor materials for lithium being consumed by battery companies [5,6]. This has resulted in a sharp increase in the price of Li_2CO_3 through the first decade of this century (Figure 4.2a) [7]. It is also predicted that lithium supplies may run out shortly due to the pervasive adoption of LIBs for certain applications (Figure 4.2b).

Sodium-ion batteries (NIBs) are becoming more popular and have the potential for large-scale grid energy storage applications due to their wide availability and low cost. Furthermore, as sodium (Na) is very abundant (the fourth most abundant element in the earth's crust), NIBs might provide an alternative chemistry to LIBs and could become more economical than LIBs in certain markets. In terms of prospective applications, NIB is anticipated to be competitive with LIB, not in terms of thermodynamic reasons but somewhat in terms of kinetic reasons [8]. There are certain cases where electrochemical performance of NIBs has been proven better than that of LIBs [9].

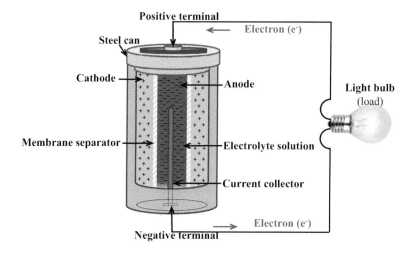

FIGURE 4.1 A battery and its components

Magnesium-ion (Mg-ion) batteries have been long considered to be promising rechargeable batteries due to their high energy density that originates from the bivalency of magnesium. Furthermore, magnesium is benign and is the fifth most abundant element in the earth's crust (13.9% as compared to $7 \times 10^{-4}\%$ for lithium). The working principle of a Mg-ion battery is similar to that of the LIB; in brief, an anode reaction involves electrochemical magnesium deposition\dissolution, whereas the cathode reaction follows cation insertion. However, in the past 20 years, this proved to work better for LIBs, whereas improvement in Mg-ion systems did not make it to the manufacturing line.

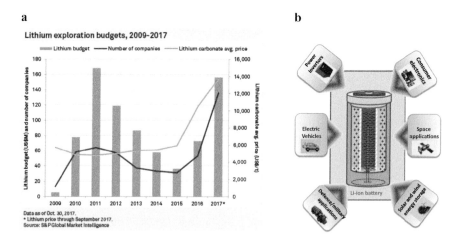

FIGURE 4.2 (a) Lithium exploration budget (2009–2017), increase in price of lithium, lithium carbonate. (b) Lithium-ion battery applications.

The metal-sulfur battery is one of the most promising next-generation rechargeable batteries. The sulfur (S) batteries offer an economical, earth-abundant cathode component that offers up to a fivefold increase in energy density compared with present LIBs. Investigators have considered sulfur as a cathode since the 1960s; however, recently numerous developments have been made in making sustainable composites that can be used for commercial applications. The electrochemical coupling of metal and sulfur is responsible for the high energy of the metal-sulfur battery. Multiple electron transfer reactions occur in the sulfur cathode during cycling; therefore, it possesses a high theoretical capacity of 1672 mA h g^{-1}, which is an order of magnitude more advanced than that of the transition-metal oxide cathodes [10]. The Li-S battery has been considered the most promising among all succeeding battery technologies. However, Na-S and Mg-S battery technologies are slowly progressing to replace Li-S batteries.

Metal-air batteries with conversion chemistry are well thought out as a favorable candidate to take over metal-ion batteries as they hold high energy density [11]. In metal-air batteries, electricity generation occurs through the redox reaction between metal and oxygen in air (O_2). They appear as open-cell structures with the cathode active material and oxygen, which is continuously and almost infinitely required from an external source (air) [12]. In case of metal-air batteries, various metals such as Zn, Ca, Fe, Cd, and Al in aqueous electrolyte and Li in nonaqueous electrolyte have been studied. Specifically, the Zn-air battery has potential for use as an alternative energy storage system. The theoretical specific energy density of Zn-air batteries is 1084 W h kg^{-1}, whereas for Li-air batteries, it is 5200 W h kg^{-1} when the mass of oxygen is included.

Carbon nanomaterials are extensively used in rechargeable batteries due to their high electrical conductivity and stability. Carbon has various allotropes composed of sp^2-conjugated carbon atoms, and their microscopic architectures and high surface areas make them the appropriate choice for rechargeable battery materials. Among all carbon nanostructures, starting from classical lithium intercalation in graphite to more advanced carbon nanoparticles, nanotubes and graphene have been studied in the past few years as potential anode materials in rechargeable batteries [13]. This chapter describes the various strategies used for the rational design of carbonaceous materials to address the function of rechargeable batteries and their role. More importantly, the role of carbonaceous materials in future development of next-generation rechargeable batteries is discussed.

4.2 Li-ION BATTERIES

In LIBs, lithium ions are shuttled between cathode and anode (two host electrodes) during the charge/discharge processes. A typical LIB consists of a graphite negative electrode (anode), a non-aqueous liquid electrolyte and a positive electrode (cathode) (Figure 4.3).

LIBs have been outperforming other existing battery technologies. They account for 63% of worldwide sales among all rechargeable batteries [14]. Lithium metal as the anode material in battery technology has several advantages. LIB has been considered the high-energy storage system (Figure 4.4) because of the fact that lithium is the

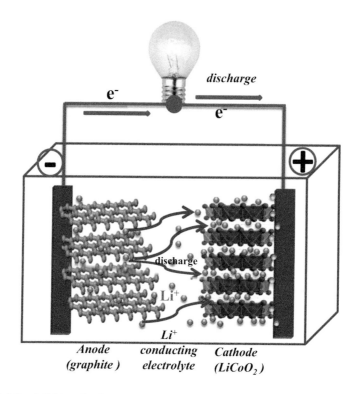

FIGURE 4.3 A lithium-ion battery (LIB). Negative electrode (graphite), positive electrode ($LiCoO_2$), separated by a non-aqueous liquid electrolyte.

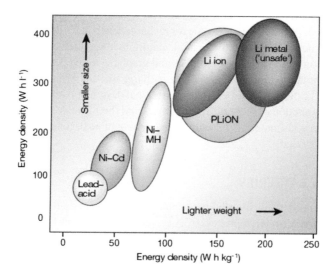

FIGURE 4.4 Comparison of the different battery technologies in terms of gravimetric energy density. (Reproduced with permission from J.M. Tarascon and M. Armand, *Nature* 414, 2001, 359–367.)

most electropositive (−3.04 V versus standard hydrogen electrode) material and the lightest (equivalent weight M = 46.94 g mol^{-1}, and specific gravity ρ = 40.53 g cm^{-3}) metal. Since the birth of LIBs in the 1980s, several carbon nanomaterials have been studied as an anode of the LIBS, although only graphitic carbon as an anode material is still the leading material among the commercially available LIBs. Other carbon nanostructures such as two-dimensional (2D) graphene and one-dimensional (1D) carbon nanotubes (CNTs) have gained attention in LIBs due to their irreplaceable structures and high mechanical, electrical and electrochemical properties. However, $LiMO_2$ (where M is Co, Ni or Mn), $LiMn_2O_4$ and $LiFePO_4$ would propose that the family of compounds are being used as cathode material in today's batteries [15–17].

4.2.1 Cathode Material

4.2.1.1 $LiCoO_2$

John B. Goodenough and his research group first used lithium cobalt oxide ($LiCoO_2$) as cathode material in 1980. $LiCoO_2$, having a similar structure to dichalcogenides, shows good electrochemical extraction of lithium and thus would be suitable for cathode material in LIBs [15]. Nevertheless, the theoretical capacity of the $LiCoO_2$ is comparatively low [14] at 140 mA h g^{-1}, because only around 0.5 Li/Co can be reversibly cycled without producing cell capacity damage due to it experiencing phase change during lithium intercalation and deintercalation [18]. Another important feature of Li_xCoO_2 is that its electrical conductivity dramatically changes with composition. A commercially available battery thus commonly uses poly(vinylidene fluoride) (PVDF) as a binder and conductive carbon to enhance the electrical conductivity [17].

4.2.1.2 $LiMn_2O_4$

Lithium manganese oxide (LMO, $LiMn_2O_4$) of spinel structure is a very propitious material for its prospects of offering not only a low-cost but also an environmentally benign cathode material, originally proposed by Thackeray et al. [16]. The $LiMn_2O_4$ spinel is the focus of much interest as a cathode of a high-power lithium battery for hybrid electric vehicles (HEVs), although under high drain rates its capacity is only around 80 mA/g [4]. This cathode material has been overwhelmed by self-discharge when held at full charge, especially at higher temperatures, which is mainly due to the manganese dissolution into the electrolyte and Jahn-Teller distortion. To overcome this problem, a conductive additive to $LiMn_2O_4$ has been considered recently to increase the conductivity of the hybrid material. The hybrids of nano-$LiMn_2O_4$ attached on graphene and CNTs have shown great improvement in the performance of LIBs due to the small particle size of active material and improved electrical conductivity of the electrode [19–24].

4.2.1.3 $LiFePO_4$

The olivine-type structure $LiFePO_4$ has become a new vital material that possesses high-performance cathode material for next-generation LIBs [25]. However, due to the low electrical conductivity and limited lithium solid-state diffusion, its performance in the battery is restricted. Nanocarbon as a conductive agent typically helps to improve the electrical conductivity while increasing the material porosity in which the solid-state diffusion distances are considerably shortened. Carbon nanomaterials including graphite, carbon black, acetylene black, nanotubes, graphene [26–30], etc. have been used for the preparation of a nanocarbon/LiFePO$_4$ composite. Recently, a study showed that LiFePO$_4$ with wrapped reduced graphene oxide sheets composites cathode showed a specific capacity of 70 mA h g^{-1} at 60C discharge rate and exhibited good capacity and decay rate up to 1,000 cycles [31]. Although LiFePO$_4$/nanocarbon materials have been considered one of the best cathode materials for LIBs, the role of carbon is to increase electrical conductivity between LiFePO$_4$ particles and it is still not clear that carbon coating is actually refining the lithium diffusion in the electrode material.

4.2.2 Anode Material

4.2.2.1 Graphite

Graphite is a widely used anode material in rechargeable LIBs due to its high Coulombic efficiency, which is the ratio of the extracted lithium to the inserted lithium [14]. The intercalation compound (thermodynamic equilibrium saturation concentration) of graphite is LIC$_6$, where the six-carbon atom can hold only one lithium, and thus the specific capacity of graphite-based LIBs is relatively low (theoretical value: 372 mA h g^{-1}) [32]. Therefore, the relatively low capacity of graphite limits its application in next-generation LIBs with high energies. To achieve the high-energy density, new anode materials with huge capacity and high recharge rate are needed. Graphene, a 2D honeycomb lattice, has been studied as an anode material as it exhibits high capacity and good rate performance, although it has low coulombic efficiency [31]. Metallic lithium deposited in the graphitic anode causes safety issues. Therefore, it is highly recommended that the anode material for advanced LIBs should have a higher reversible capacity and superior charge/discharge rate than graphite.

4.3 SODIUM-ION BATTERIES

LIBs have remarkable potential to meet the challenges of renewable energy development and serve the market requirement for electric vehicles, but the abundance of lithium in the earth's crust is limited to only 20 ppm [33]. Certainly, the material costs (mainly Li$_2$CO$_3$) for LIBs have been constantly increasing during the first decade of this century. In contrast, sodium (Na) is not only the most abundant element in the earth's crust, but it also is the second-lightest and smallest alkali metal in the periodic table. The sodium-ion battery (NIB) is therefore the correct alternative to LIB with respect to standard electrode potential and abundance.

NIBs and LIBs are the same in structure, components, systems, and charge storage mechanisms except that the lithium ions are replaced with sodium ions. In case of NIBs, sodium insertion materials with aprotic solvent as electrolyte are required, and obviously, they must be free from metallic sodium because it may cause failure of the batteries. Carbon nanomaterials including expanded graphite, graphene, carbon nanospheres, carbon nanofibers, carbon nanotubes, carbon nanosheets and porous carbons as anode are the preferable choices for NIBs due to their low potential, high capacity, and abundance, similar to LIBs [34,35].

Recently, graphene, doped graphene and graphene nanocomposites have emerged as anode materials for NIBs because of their unique 2D structures, mechanical strength, high electrical conductivity, and large surface area [36]. Reduced graphene oxide (rGO) and N- and S-doped rGO show excellent performance as NIB anode materials due to synergistic effect by –N and –S and its mesoporous structure [37]. Ma et al. recently reported highly doped graphene nanosheets with tunable N and S codopants (NS-GNS) electrode demonstrates a high reversible capacity of around 400 mA h g^{-1} and ultrahigh life cycle [38]. Graphene-based composites such as the three-phase storage mechanism in $Li_4Ti_5O_{12}$ anodes for room-temperature NIBs and a zero-strain layered metal oxide, that is, P2-type $Na_{0.66}$ [$Li_{0.22}Ti_{0.78}$]O_2, have produced strong concerns. As typical Ti-based compounds, the structures of TiO_2 and NASICON-type $NaTi_2(PO_4)_3$ were used with graphene for the NIB anode [39]. Graphene with several oxide nanostructures has drawn great attention as anode material for NIBs. Recently, Fe_2O_3@GNS, vanadium oxide, $FeWO_4$/graphene composites have been used in conversion reactions for sodium cyclability [37].

4.4 MAGNESIUM-ION BATTERIES

Rechargeable magnesium-ion batteries (MIBs) have become the focus of recent research and technology development. The abundance of magnesium in the earth's crust is approximately 10^4 times that of lithium. The thermodynamics properties of magnesium and its incorporation into electrode materials and cost effectiveness make it popular for use in rechargeable batteries [40,41]. MIBs are well received by the scientific community as they are an environmentally friendly, non-toxic, good alternative to the Li-ion systems due to their high volumetric capacity (3833 mA h/ cc) [42]. Although MIBs hold some positive attributes, the development of MIB technology is slow to progress along with that of lithium-ion technology due to the need for development of a suitable electrolyte that will help to reverse the release of Mg^{2+} ions from a magnesium metal anode [40]. In case of MIBs, the chevrel phases, $M_xMO_6T_8$ (M = metal, T = S, Se), are of great interest due to their remarkable electromagnetic, thermoelectric and catalytic properties [43]. Graphene, defective graphene and graphene-based composites have been reported recently to be used as high-capacity anode materials for Mg-ion batteries [44–46]. Shenoy and co-workers reported that magnesium storage capacity can be achieved as high as 1042 mA h g^{-1} in graphene with 25% divacancy defects [45].

4.5 METAL-SULFUR BATTERIES

Sulfur is one of the most abundant elements on earth. It is an electrochemically active material and can accept up to two electrons per atom at ~2.1 V versus Li/ Li$^+$. Conventional metal-sulfur cells consist of a metal (e.g. Li, Mg, etc.) anode, a sulfur composite cathode and an organic liquid electrolyte. A typical charge-discharge profile for the first cycle of Li-S cells is shown in Figure 4.5 [47]. The sulfur composite cathode material in metal-sulfur batteries has a high theoretical capacity of 1675 mA h g^{-1}, and Li-S batteries have a theoretical energy density of ~2600 W h kg^{-1}. However, sulfur undergoes structural changes during cycling and

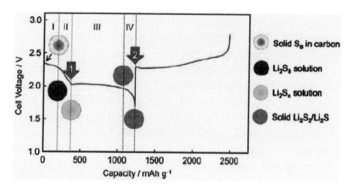

FIGURE 4.5 A typical discharge and charge voltage profile of the first cycle of Li/S cells. (Reproduced with permission from S.S. Zhang, *J. Power Sources* 231, 2013, 153–162.)

produces soluble polysulfides and insoluble sulfides. Therefore, metal-sulfur batteries suffer from life cycle and system inefficiency [48]. A number of procedures have been developed to form sulfur composites with favourable structures and properties to enhance the discharge capacity, Coulombic efficiency and cyclability. In recent years, porous hollow carbon@sulfur, graphene-wrapped sulfur particles, sulfur-impregnated disordered carbon nanotubes and activated carbon-sulfur have been prepared as electrode material for the Li-S battery [49–53]. A combination of a magnesium anode with a sulfur cathode is also a proficient electrochemical coupling due to the advantages of safety, low cost and high theoretical energy density of over 3200 Wh L^{-1}. Vinayan et al. reported a Mg-S battery consisting of a graphene-sulfur nanocomposite as the cathode, a Mg-carbon composite as the anode and a non-nucleophilic magnesium-based complex in tetraglyme solvent as the electrolyte [54], which delivered higher capacity (448 mA h g^{-1}) and good cyclability (236 mA h g^{-1} after the 50th cycle). Qiu et al. reported Li/S@ N-doped graphene (60% S) with specific discharge capacity of 978 mA h g^{-1} at 0.2C after 150 cycles [55]. In another example, Wang et al. used N-doped graphene/sulfur (NGS) composites with 80 wt% of sulfur that possessed a reversible capacity of 1356.8 mA h g^{-1} at 0.1C and long cycle stability of 578.5 mA h g^{-1} at 1C after 500 cycles [56].

4.6 METAL-AIR BATTERIES

The important criteria of a metal-air battery are the combination of a metal anode with high-energy density and an air electrode with open structure for cathode active materials (i.e. oxygen) from air, thus exhibiting higher theoretical energy density than other secondary batteries. The electricity of a metal-air battery is generated through a redox reaction between metal and oxygen in air [11]. There are several metal-air chemistries found in the literature. Among them, the primary Zn-air cell has been commercialized; Al-air and Mg-air cells have found military applications and are under development as saline systems [11]. New research interest has been found in Li-air batteries that are purely based on Li-intercalation electrodes. The important task of state-of-the-art metal-air batteries includes first elucidating the basic

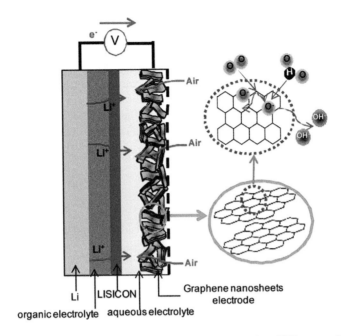

FIGURE 4.6 Structure of the rechargeable Li-air battery based on GNSs as an air electrode. (Reproduced with permission from E. Yoo and H. Zhou, *ACS Nano* 5(4), 2011, 3020–3026.)

oxygen chemistry and then exploiting highly efficient air electrodes for accelerating the electrochemical oxygen reduction reaction (ORR) and/or oxygen evolution reaction (OER). In this context, researchers have been dedicated to searching for the bifunctional electrocatalyst for ORR/OER and the electrode structures for facilitating mass/charge transport and interface interaction.

Yoo et al. recently developed a metal-free graphene catalyst for the rechargeable Li-air battery (Figure 4.6), which showed a high discharge voltage that was closer to 20 wt% Pt/carbon black [57]. Zhang et al. reported an efficient Zn-air battery with three-dimensional (3D) mesoporous carbon foams co-doped with N and P, simply by pyrolysing polyaniline aerogels obtained from a template-free polymerisation of aniline in the presence of phytic acid [58]. Another report by Zhang et al. included a long-life Li-air cell produced from cross-linked network gel (CNG) consisting of single-walled carbon nanotubes (SWCNTs) and ionic liquid [59]. This Li-air cell has sustained repeated cycling in ambient air for 100 cycles for the duration of 78 days, with discharge capacity of 2000 mA h g^{-1}. Tetragonal $CoMn_2O_4$ spinel nanoparticles grown on graphene can act as a bifunctional catalyst for Li-air batteries, as has been reported by Wang et al. [60].

4.7 SUMMARY AND OUTLOOK

In this chapter, we summarised the nanocarbons including graphene, graphite, carbon nanotube, mesoporous carbon, carbon nanofiber and activated carbon and their composites as electrode materials for battery applications. Furthermore, different types of batteries, such as LIBs, NIBs, MIBs, metal-sulfur and metal-air batteries, were critically studied. The role of nanocarbon materials and their derivatives as cathode as well as anode materials to the different types of batteries have been explored. Different carbon nanomaterials and techniques have paved the way for the design and fabrication of the materials to the electrode in order to achieve the best performances as storage materials. It is worth mentioning here that the performance of these batteries needs to be manipulated substantially to achieve the requirements of high energy and power density for future energy storage systems. Careful engineering of the structure and electrical conductivity of the material can further improve the capacity and rate capability of the device.

REFERENCES

1. N.A. Kaskhedikar, J. Maier, Lithium storage in carbon nanostructures, *Adv. Mater.* 21, 2009, 2664–2680. doi: 10.1002/adma.200901079
2. Schafhaeutl, Ueber weisses und graues Roheisen, Graphit-Bildung u. s. w., *J. Für Prakt. Chemie.* 76, 1859, 257–310. doi: 10.1002/prac.18590760151
3. D.R. Tobergte, S. Curtis, Ueber die Verbindungen des Kohlenstoffes mit Silicium, *J. Chem. Inf. Model.* 53, 2013, 1689–1699. doi: 10.1017/CBO9781107415324.004
4. M.S. Whittingham, Lithium batteries and cathode materials, *Chem. Rev.* 104, 2004, 4271–4302. doi: 10.1021/cr020731c
5. M.D. Slater, D. Kim, E. Lee, C.S. Johnson, Sodium-ion batteries, *Adv. Funct. Mater.* 23, 2013, 947–958. doi: 10.1002/adfm.201200691
6. 2012 U.S. Geological Survey, Reston, VA, "Lithium" in Mineral Commodity Summaries, 2012.
7. "Market," The Lithium Site, 2012 (Accessed February 2012).
8. S.P. Ong, V.L. Chevrier, G. Hautier, A. Jain, C. Moore, S. Kim, X. Ma, G. Ceder, Voltage, stability and diffusion barrier differences between sodium-ion and lithium-ion intercalation materials, *Energy Environ. Sci.* 4, 2011, 3680. doi: 10.1039/c1ee01782a
9. C. Zhu, X. Mu, P.A. Vanaken, Y. Yu, J. Maier, Single-layered ultrasmall nanoplates of MoS2 embedded in carbon nanofibers with excellent electrochemical performance for lithium and sodium storage, *Angew. Chemie.* 53, 2014, 2152–2156. doi: 10.1002/anie.201308354
10. J. Lochala, D. Liu, B. Wu, C. Robinson, J. Xiao, Research progress toward the practical applications of lithium-sulfur batteries, *ACS Appl. Mater. Interfaces* 9, 2017, 24407–24421. doi: 10.1021/acsami.7b06208
11. F. Cheng, J. Chen, Metal–air batteries: From oxygen reduction electrochemistry to cathode catalysts, *Chem. Soc. Rev.* 41, 2012, 2172. doi: 10.1039/c1cs15228a
12. J.S. Lee, S.T. Kim, R. Cao, N.S. Choi, M. Liu, K.T. Lee, J. Cho, Metal-air batteries with high energy density: Li-air versus Zn-air, *Adv. Energy Mater.* 1, 2011, 34–50. doi: 10.1002/aenm.201000010
13. L. Dai, D.W. Chang, J.-B. Baek, W. Lu, Carbon nanomaterials for advanced energy conversion and storage, *Small* 8, 2012, 1130–66. doi: 10.1002/smll.201101594
14. J.M. Tarascon, M. Armand, Issues and challenges facing rechargeable lithium batteries., *Nature* 414, 2001, 359–367. doi: 10.1038/35104644
15. K. Mizushima, P.C. Jones, P.J. Wiseman, J.B. Goodenough, LixCoO2 ($0<x<-1$): A new cathode material for batteries of high energy density, *Mater. Res. Bull.* 15, 1980, 783–789. doi: 10.1016/0025-5408(80)90012-4
16. M.M. Thackeray, W.I.F. David, P.G. Bruce, J.B. Goodenough, Lithium insertion into manganese spinels, *Mater. Res. Bull.* 18, 1983, 461–472. doi: 10.1016/0025-5408(83)90138-1

17. M. Park, X. Zhang, M. Chung, G.B. Less, A.M. Sastry, A review of conduction phenomena in Li-ion batteries, *J. Power Sources* 195, 2010, 7904–7929. doi: 10.1016/j.jpowsour.2010.06.060
18. J.N. Reimers, Electrochemical and in situ X-ray diffraction studies of lithium intercalation in Li[sub x]CoO[sub 2], *J. Electrochem. Soc.* 139, 1992, 2091. doi: 10.1149/1.2221184
19. B. Lin, Q. Yin, H. Hu, F. Lu, H. Xia, $LiMn_2O_4$ nanoparticles anchored on graphene nanosheets as high-performance cathode material for lithium-ion batteries, *J. Solid State Chem.* 209, 2014, 23–28. doi: 10.1016/j.jssc.2013.10.016
20. H. Xu, B. Cheng, Y. Wang, L. Zheng, X. Duan, L. Wang, J. Yang, Y. Qian, Improved electrochemical performance of $LiMn_2O_4$/graphene composite as cathode material for lithium ion battery, *Int. J. Electrochem. Sci.* 7, 2012, 10627–10632.
21. K.C. Jiang, X.L. Wu, Y.X. Yin, J.S. Lee, J. Kim, Y.G. Guo, Superior hybrid cathode material containing lithium-excess layered material and graphene for lithium-ion batteries, *ACS Appl. Mater. Interfaces* 4, 2012, 4858–4863. doi: 10.1021/am301202a
22. J. Li, X. Zhang, R. Peng, Y. Huang, L. Guo, Y. Qi, $LiMn_2O_4$ /graphene composites as cathodes with enhanced electrochemical performance for lithium-ion capacitors, *RSC Adv.* 6, 2016, 54866–54873. doi: 10.1039/c6ra09103b
23. Y.Q. Huayun Xu, Bin Cheng, Yunpo Wang, Long Zheng, Xinhui Duan, Lihui Wang, Jian Yang, A study on improving drying performance of spinel type $LiMn_2O_4$ as a cathode material for lithium ion battery, *Int. J. Electrochem. Sci.* 6, 2011, 5462–5469.
24. H. Xia, Z. Luo, J. Xie, Nanostructured $LiMn_2O_4$ and their composites as high-performance cathodes for lithium-ion batteries, *Prog. Nat. Sci. Mater. Int.* 22, 2012, 572–584. doi: 10.1016/j.pnsc.2012.11.014
25. A. Eftekhari, $LiFePO_4$/C nanocomposites for lithium-ion batteries, *J. Power Sources* 343, 2017, 395–411. doi: 10.1016/j.jpowsour.2017.01.080
26. H.C. Shin, W. Il Cho, H. Jang, Electrochemical properties of carbon-coated $LiFePO_4$ cathode using graphite, carbon black, and acetylene black, *Electrochim. Acta* 52, 2006, 1472–1476. doi: 10.1016/j.electacta.2006.01.078
27. D. Zhao, Y.L. Feng, Y.G. Wang, Y.Y. Xia, Electrochemical performance comparison of $LiFePO_4$ supported by various carbon materials, *Electrochim. Acta* 88, 2013, 632–638. doi: 10.1016/j.electacta.2012.10.101
28. G. Qin, Q. Wu, J. Zhao, Q. Ma, C. Wang, C/$LiFePO_4$/multi-walled carbon nanotube cathode material with enhanced electrochemical performance for lithium-ion batteries, *J. Power Sources* 248, 2014, 588–595. doi: 10.1016/j.jpowsour.2013.06.070
29. H. Wu, Q. Liu, S. Guo, Composites of graphene and $LiFePO_4$ as cathode materials for lithium-ion battery: A mini-review, *Nano-Micro Lett.* 6, 2014, 316–326. doi: 10.1007/s40820-014-0004-6
30. Y. Shi, S.-L. Chou, J.-Z. Wang, D. Wexler, H.-J. Li, H.-K. Liu, Y. Wu, Graphene wrapped $LiFePO_4$/C composites as cathode materials for Li-ion batteries with enhanced rate capability, *J. Mater. Chem.* 22, 2012, 16465. doi: 10.1039/c2jm32649c
31. X. Zhou, F. Wang, Y. Zhu, Z. Liu, Graphene modified $LiFePO_4$ cathode materials for high power lithium ion batteries, *J. Mater. Chem.* 21, 2011, 3353. doi: 10.1039/c0jm03287e
32. M. Liang, L. Zhi, Graphene-based electrode materials for rechargeable lithium batteries, *J. Mater. Chem.* 19, 2009, 5871. doi: 10.1039/b901551e
33. N. Yabuuchi, K. Kubota, M. Dahbi, S. Komaba, Research development on sodium-ion batteries, *Chem. Rev.* 114, 2014, 11636–11682. doi: 10.1021/cr500192f
34. W. Luo, F. Shen, C. Bommier, H. Zhu, X. Ji, L. Hu, Na-ion battery anodes: Materials and electrochemistry, *Acc. Chem. Res.* 49, 2016, 231–240. doi: 10.1021/acs.accounts.5b00482
35. H. Hou, X. Qiu, W. Wei, Y. Zhang, X. Ji, Carbon anode materials for advanced sodium-ion batteries, *Adv. Energy Mater.* 201602898, 2017, 1–30. doi: 10.1002/aenm.201602898
36. Y. Liang, W.H. Lai, Z. Miao, S.L. Chou, Nanocomposite materials for the sodium–ion battery: A review, *Small* 14, 2018, 1702514. doi: 10.1002/smll.201702514

37. S. Wu, R. Ge, M. Lu, R. Xu, Z. Zhang, Graphene-based nano-materials for lithium-sulfur battery and sodium-ion battery, *Nano Energy* 15, 2015, 379–405. doi: 10.1016/j.nanoen.2015.04.032
38. Y. Ma, Q. Guo, M. Yang, Y. Wang, T. Chen, Q. Chen, X. Zhu, Q. Xia, S. Li, H. Xia, Highly doped graphene with multi-dopants for high-capacity and ultrastable sodium-ion batteries, *Energy Storage Mater.* 13, 2018, 134–141. doi: 10.1016/j.ensm.2018.01.005
39. Y. Sun, L. Zhao, H. Pan, X. Lu, L. Gu, Y.S. Hu, H. Li, et al., Direct atomic-scale confirmation of three-phase storage mechanism in Li4Ti5O12 anodes for room-temperature sodium-ion batteries, *Nat. Commun.* 4, 2013, 1810–1870. doi: 10.1038/ncomms2878
40. M.M. Huie, D.C. Bock, E.S. Takeuchi, A.C. Marschilok, K.J. Takeuchi, Cathode materials for magnesium and magnesium-ion based batteries, *Coord. Chem. Rev.* 287, 2015, 15–27. doi: 10.1016/j.ccr.2014.11.005
41. D. Aurbach, Z. Lu, A. Schechter, Y. Gofer, H. Gizbar, R. Turgeman, Y. Cohen, M. Moshkovich, E. Levi, Prototype systems for rechargeable magnesium batteries, *Nature* 407, 2000, 724–727. doi: 10.1038/35037553
42. P. Saha, M.K. Datta, O.I. Velikokhatnyi, A. Manivannan, D. Alman, P.N. Kumta, Rechargeable magnesium battery: Current status and key challenges for the future, *Prog. Mater. Sci.* 66, 2014, 1–86. doi: 10.1016/j.pmatsci.2014.04.001
43. D. Aurbach, G.S. Suresh, E. Levi, A. Mitelman, O. Mizrahi, O. Chusid, M. Brunelli, Progress in rechargeable magnesium battery technology, *Adv. Mater.* 19, 2007, 4260–4267. doi: 10.1002/adma.200701495
44. E. Sheha, M.H. Makled, W.M. Nouman, A. Bassyouni, S. Yaghmour, S. Abo-Elhassan, Vanadium oxide/graphene nanoplatelet as a cathode material for Mg-Ion battery, *Graphene* 05, 2016, 178–188. doi: 10.4236/graphene.2016.54015
45. D. Er, E. Detsi, H. Kumar, V.B. Shenoy, Defective graphene and graphene allotropes as high-capacity anode materials for Mg ion batteries, *ACS Energy Lett.* 1, 2016, 638–645. doi: 10.1021/acsenergylett.6b00308
46. J. Zhu, Y. Ren, B. Yang, W. Chen, J. Ding, Embedded Si/graphene composite fabricated by magnesium-thermal reduction as anode material for lithium-ion batteries, *Nanoscale Res. Lett.* 12, 2017, 627–634. doi: 10.1186/s11671-017-2400-6
47. S.S. Zhang, 156 Liquid electrolyte lithium/sulfur battery: Fundamental chemistry, problems, and solutions, *J. Power Sources* 231, 2013, 153–162. doi: 10.1016/j.jpowsour.2012.12.102
48. A. Manthiram, Y. Fu, Y.S. Su, Challenges and prospects of lithium-sulfur batteries, *Acc. Chem. Res.* 46, 2013, 1125–1134. doi: 10.1021/ar300179v
49. H. Wang, Y. Yang, Y. Liang, J.T. Robinson, Y. Li, A. Jackson, Y. Cui, H. Dai, Graphene-wrapped sulfur particles as a rechargeable lithium-sulfur battery cathode material with high capacity and cycling stability, *Nano Lett.* 11, 2011, 2644–2647. doi: 10.1021/nl200658a
50. R. Elazari, G. Salitra, A. Garsuch, A. Panchenko, D. Aurbach, Sulfur-impregnated activated carbon fiber cloth as a binder-free cathode for rechargeable Li-S batteries, *Adv. Mater.* 23, 2011, 5641–5644. doi: 10.1002/adma.201103274
51. D. Aurbach, E. Pollak, R. Elazari, G. Salitra, C.S. Kelley, J. Affinito, On the surface chemical aspects of very high energy density, rechargeable Li–Sulfur batteries, *J. Electrochem. Soc.* 156, 2009, A694. doi: 10.1149/1.3148721
52. A. Manthiram, S.-H. Chung, C. Zu, Lithium-sulfur batteries: Progress and prospects, *Adv. Mater.* 27, 2015, 1980–2006. doi: 10.1002/adma.201405115
53. N. Jayaprakash, J. Shen, S.S. Moganty, A. Corona, L.A. Archer, Porous hollow carbon@sulfur composites for high-power lithium-sulfur batteries, *Angew. Chemie, Int. Ed.* 50, 2011, 5904–5908. doi: 10.1002/anie.201100637

54. B.P. Vinayan, Z. Zhao-Karger, T. Diemant, V.S.K. Chakravadhanula, N.I. Schwarzburger, M.A. Cambaz, R.J. Behm, C. Kübel, M. Fichtner, Performance study of magnesium–sulfur battery using a graphene based sulfur composite cathode electrode and a non-nucleophilic Mg electrolyte, *Nanoscale* 8, 2016, 3296–3306. doi: 10.1039/C5NR04383B
55. Y. Qiu, W. Li, W. Zhao, G. Li, Y. Hou, M. Liu, L. Zhou, et al., High-rate, ultralong cycle-life lithium/sulfur batteries enabled by nitrogen-doped graphene, *Nano Lett.* 14, 2014, 4821–4827. doi: 10.1021/nl5020475
56. C. Wang, K. Su, W. Wan, H. Guo, H. Zhou, J. Chen, X. Zhang, Y. Huang, High sulfur loading composite wrapped by 3D nitrogen-doped graphene as a cathode material for lithium–sulfur batteries, *J. Mater. Chem. A* 2, 2014, 5018–5023. doi: 10.1039/C3TA14921H.
57. E. Yoo, H. Zhou, Li-air rechargeable battery based on metal-free graphene nanosheet catalysts, *ACS Nano* 5, 2011, 3020–3026. doi: 10.1021/nn200084u
58. J. Zhang, Z. Zhao, Z. Xia, L. Dai, A metal-free bifunctional electrocatalyst for oxygen reduction and oxygen evolution reactions, *Nat. Nanotechnol.* 10, 2015, 444–452. doi: 10.1038/nnano.2015.48
59. T. Zhang, H. Zhou, A reversible long-life lithium-air battery in ambient air, *Nat. Commun.* 4, 2013, 1817. doi: 10.1038/ncomms2855
60. L. Wang, X. Zhao, Y. Lu, M. Xu, D. Zhang, R.S. Ruoff, K.J. Stevenson, J.B. Goodenough, $CoMn_2O_4$ spinel nanoparticles grown on graphene as bifunctional catalyst for lithium-air batteries, *J. Electrochem. Soc.* 158, 2011, A1379. doi: 10.1149/2.068112jes

5 Supercapacitor

5.1 INTRODUCTION

Electrochemical capacitors (ECs), also called supercapacitors (SCs), constitute a subclass of contemporary electrochemical energy storage devices that has emerged with the potential to greatly influence the path of research in the field of energy storage. An SC is a culmination of two very distinct energy storage systems: an electrical capacitor and a battery. They are designed to be fast-charging energy storage devices with intermittent energy density, which means they can deliver quick bursts of energy during peak power demands and then quickly store energy and capture excess power that is otherwise lost. SCs are power-rich energy components for energy alternating applications, for example, transportation, intermittent generators and smart grids.

A supercapacitor differs from a conventional capacitor both in the mechanism of charge storage and the fact that it has very high capacitance. A capacitor stores energy by means of a static charge as opposed to an electrochemical reaction in an SC. Applying a voltage differential on the positive and negative plates charges the capacitor, while in an SC, charge is stored by a phenomenon known as 'electrical double layer' (EDL). It operates on the simple mechanism of adsorption of electrolyte ions at the EDL on a high-surface-area electrode.

Broadly, capacitors can be divided into three types based on their charge storage mechanism. The various components and assembling of the different types of capacitors are summarised in Figure 5.1. The most basic type is the electrostatic capacitor, with a dry separator. It has very low capacitance and is mainly used for filtering signals and tuning radio frequencies (RFs). The size ranges from a few picofarads to low microfarads. The next type is the electrolytic capacitor, which is used for the filtering, buffering and coupling of power. Rated in microfarads, this capacitor has several thousand times the storage capacity of the electrostatic capacitor and uses a moist separator. The final type is the SC, rated in farads, which is thousands of times higher than the electrolytic capacitor.

What separates an SC from a battery is not only how the energy is stored but also its distinct application range. From a general point of view, lithium-ion (Li-ion) batteries can store high energy densities (up to 190 W h kg^{-1} for commercial products) with low power densities (up to about 2 kW kg^{-1}). Electrical double-layer capacitors (EDLCs) can deliver very high power density (15 kW kg^{-1}) with lower stored energy (5 W h kg^{-1}) than that of batteries. For longer charge time (lower current density), the battery can store at least 20 times more energy than an SC. But when the charge time is less (higher current density), the energy density of the battery decreases, while the energy density of the SC remains almost stable. For a charge time of a few seconds, SCs can store more energy than a battery. Applications requiring a supply of energy during short bursts (power) are better addressed by SCs, while batteries are suitable for applications requiring longer supplies of energy.

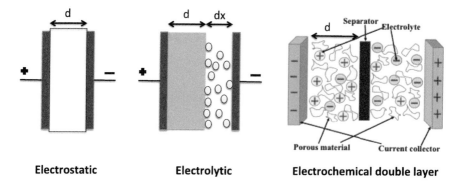

FIGURE 5.1 Various components and assembling of different types of capacitors.

Although the storage of electrical charge at the interface of a metal and an electrolyte (electrolytic capacitors) has been studied since the nineteenth century, SCs first came into existence in 1957 when H. Becker first proposed a 'low voltage electrolytic capacitor with porous carbon electrode' [1]. He believed that the energy was stored in the form of a charge in the carbon pores, similar to electrolytic capacitors. With electrodes based on 'fired tar lampblack' (according to the patent description), significantly higher capacitance values have been obtained in comparison with an electrolytic capacitor of comparable size at that time. By replacing the dielectric films from the anode aluminium foil and cathode aluminium foil surfaces of an electrolytic capacitor with activated (porous) carbon-based layers, another advance in SC research and development was realised.

Since then, carbon materials have been remarkable in revolutionising SC research. Here, we specifically discuss the evolution of carbon-based materials in the field of supercapacitor applications. Research into ECs has dramatically increased since the early 1990s and has been fuelled by an emerging number of applications requiring their unique combination of properties that include high specific energy (for a capacitive device), reliability, long cycle life, high power (both charge and discharge) and high energy efficiency.

5.2 CHARGE STORAGE MECHANISM IN SUPERCAPACITORS

5.2.1 Electrical Double-Layer Capacitor

EDLCs are ECs storing the charge electrostatically by reversible adsorption of ions of the electrolyte onto electrochemically stable high specific surface area (SSA) carbonaceous active materials [2]. Charge separation occurs on polarisation at the electrode/electrolyte interface, producing what Helmholtz [3] described in 1853 as the double-layer capacitance C according to:

$$C = \xi_r \xi_0 A/d$$

$$\text{or } C/A = \xi_r \xi_0 / d$$

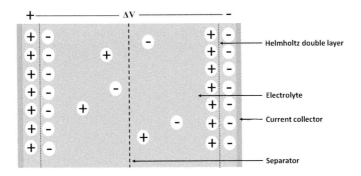

FIGURE 5.2 Schematics of an ideal electric double-layer capacitor (EDLC).

where ξ_r is the relative dielectric constant of the electrolyte, ξ_0 is the dielectric constant of vacuum, d is the effective thickness of the double layer (charge separation distance) and A is the surface area of the interface. This capacitance model was later refined by Gouy and Chapman and Stern and Geary [4] who suggested the presence of a diffuse layer in the electrolyte due to the accumulation of ions close to the electrode surface. A schematic EDLC and its charge storage mechanism are depicted in Figure 5.2.

Carbon-based materials are the most researched materials for EDLC-type SCs because of their enormous surface area and improved electrical conductivity. EDLCs have great advantages, like superior rate capability or cycling performance, but they lack in capacitance. The double-layer capacitance is between 5 and 20 $\mu F\ cm^{-2}$ depending on the electrolyte used. The capacitance thus must be increased in order to be able to achieve comparable storage capacity to that of rechargeable batteries.

5.2.2 Pseudocapacitors

Apart from the physical separation and storage of charge at the electrical double layer, sometimes some materials use fast and reversible redox reactions at the electrode surface that contribute an additional faradaic capacitance to the EDL capacitance. This phenomenon is known as pseudocapacitance. Since the capacitance is not 'electrostatic' in nature, the name *pseudocapacitance* is coined to differentiate it from non-faradaic processes. Pseudocapacitance takes place when an electrochemical charge transfer occurs to an extent limited by a finite amount of active material on the electrode surface. The fact that charge storage is based on a redox process means that this type of supercapacitor is somewhat battery-like (faradic) in its behaviour.

Until now, the most commonly explored classes of pseudocapacitive materials are the family of conducting polymers, such as polyaniline (PANI), polypyrrole (PPy) or derivatives of polythiophene (PTh) and transition metal oxides and dichalcogenides, such as RuO_2, MnO_2, MoS_2, etc. Porous carbons that possess a significant proportion of heteroatoms (typically oxygen or nitrogen) and/or surface functionalities can also contain a pseudocapacitive component in their overall capacitance. That is, the double-layer capacitance derived from the extended carbon surface is supplemented by the pseudocapacitive contributions arising from the redox active functionalities, thereby substantially increasing the total capacitance of the material.

5.2.3 HYBRID SUPERCAPACITORS

Hybrid SCs utilise both the charge storage mechanisms described previously in its electrode. Here, one electrode is made up of pure EDLC-type material involving a non-faradaic process, while the other electrode is composed of some pseudocapacitive components, such as conducting polymers that undergo the redox processes. So, this is a kind of asymmetric SC where both the electrodes differ from each other. The term *asymmetric supercapacitors* is suggested by Brousse et al. to be used for the devices with only capacitive/pseudocapacitive electrodes (e.g. MnO_2/graphene/carbon nanotubes) to avoid confusion with 'hybrid' devices in between an SC and a battery [5]. Hybrid SCs consisting of a pseudocapacitive faradaic electrode (for high energy density) and a capacitive electrode (for high power density and long cycle life) could achieve even higher energy density because the hybridisation of these two electrodes can further broaden the operating voltage and increase the capacitance of the hybrid capacitor.

A lot of research is going on in developing hybrid SCs because of the added advantage they have over just an EDLC or a pseudocapacitor. Various nanostructured materials, such as Li-intercalated compounds and transition metal (Ni, Co) oxides/hydroxides are emerging as the pseudocapacitive electrodes, while activated carbons, graphene or CNTs are commonly chosen as the capacitive electrode. Though these types of metal-based oxides/hydroxides often can undergo phase transformations in the bulk phase and suffer from poor conductivity, engineering them at the nano-scale while giving a conductive support like carbon often promotes their pseudocapacitive property and overall electrochemical performance.

5.3 DEVICE ARCHITECTURE OF SUPERCAPACITOR SYSTEM

5.3.1 SANDWICH-TYPE SUPERCAPACITOR

Sandwich-type or traditional SCs (Figure 5.3a) [6] are solid-state SC devices where the two plates or the electrodes are assembled one upon another, separated by a separator. Generally, any porous but insulating polymers are used as separators to prevent short-circuit in the device, while an ionically conducting gel electrolyte is used to carry the electrolyte ions in a horizontal ionic movement. Active materials are generally drop-cast, spin-coated or deposited on current collectors to make the electrodes. Sandwich-type SCs often suffer from the problem of short-circuit or internal resistance if the two electrodes are too close or too far apart (having air-gap) from each other. Because of its three-dimensional (3D) architecture, they often have poor volumetric performance and hence are not suitable to be used in portable microelectronics.

5.3.2 IN-PLANE SUPERCAPACITOR

Planar SCs are a more recent development in the growing field of SCs being used as a portable power source in microelectronics or personal digital assistants. In-plane SCs (Figure 5.3b) [6] have both the electrodes in the same plane, where the electrolyte ions move horizontally, and this efficiently reduces the ionic diffusion path, leading to enhanced capacitance per unit area or volume of the device. Because of the sophisticated methods that are used in developing in-plane SCs, such as laser

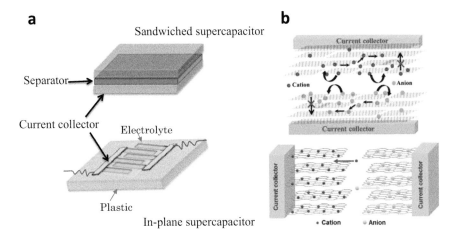

FIGURE 5.3 Schematics of (a) a sandwich and (b) an in-plane micro-supercapacitor. (Reproduced with permission from J.J. Yoo et al., *Nano Lett.* 4, 2011, 1423–1427.)

scribing or lithography, the electrodes can be fabricated in the micro scale with less than a 50-micrometre space between the two electrodes. Thus, in-plane micro-supercapacitors are currently emerging as an alternative to micro-batteries, with unprecedented volumetric energy densities that are efficient enough to power small electronic devices, such as watches or fitness bands.

5.4 KEY PARAMETERS FOR SUPERCAPACITOR PERFORMANCE EVALUATION

5.4.1 Specific Capacitance

There are several parameters to characterise a good SC device, the most important of which is the specific capacitance (C_{SP}) (areal, volumetric or gravimetric; expressed in Farads per cm^2, cm^3 or g). C_{SP} could be measured and calculated by mainly two methods, galvanostatic charge/discharge (GCD) and cyclic voltammetry (CV) measurements, by Equations 5.1 and 5.2. From GCD, C_{SP} (Equation 5.1) can be addressed as

$$C_{SP} = \frac{I \Delta t}{\Delta V} \frac{1}{\langle A|V|M \rangle} \tag{5.1}$$

where I (in A) is the discharge current, Δt (in s) is the discharge time, A (in cm^2) is the area of the electrode, V (in cm^3) is the volume of the electrode and M (in g) is the active mass of the electrode material, and ΔV (in V) is the working voltage.

Similarly, by CV technique, specific capacitance C_{SP} (Equation 5.2) can be given as

$$C_{SP} = \frac{I_{avg}}{\langle A|V|M \rangle \nu} \tag{5.2}$$

where I_{avg} is the average current obtained from cathodic and anodic sweeps and υ is the scan rate.

For both EDLC and pseudocapacitive materials, a good electrical conductivity and high surface area will contribute to high specific capacitance, as the high electrical conductivity generates a higher charge/electron carrier mobility and a high surface area promotes charge absorption or provides more redox reaction sites.

5.4.2 Energy Density and Power Density

Besides obtaining a high specific capacitance, several other parameters are also there to determine the overall performance of an SC, for example, specific energy and power density. Both energy density (E) and power density (P) are crucial to an energy storage device, as they decide how efficient a system is. As revealed in Figure 5.4 (Ragone plot for various energy storage systems), SCs offer much higher power density than various Li-ion batteries. However, their application is limited by very low energy density. Thus, one of the biggest challenges for their application is to enhance the energy density while maintaining the other inherent qualities, such as specific power and cycling stabilities. The specific energy and power can be expressed as follows by Equations 5.3 and 5.4, respectively:

$$E = \frac{\Delta V^2}{2}(C_{sp})_{\langle A|V|M \rangle} \tag{5.3}$$

$$P = \frac{E}{\Delta t} \tag{5.4}$$

where ΔV is the voltage window and Δt is the discharge time corresponding to the C_{SP} value calculated from GCD measurement. As shown from Equation 5.3, the key to enhancing E is to increase the specific capacitance and voltage window. Many modifications and optimizations, in not only active materials but also in electrolyte

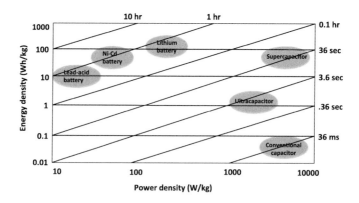

FIGURE 5.4 Ragone plot for various energy storage devices (ESDs), with diagonal discharge time.

selection and device architecture design, go towards ensuring the maximum energy density of an SC system. For example, organic electrolytes (3–4 V) provide a higher stable operating voltage window than aqueous electrolytes (1 V) and hence result into higher energy densities.

5.4.3 CYCLING STABILITY

Another determining parameter for practical applications of supercapacitors is their excellent cycling stabilities. EDLCs can withstand more than 1,00,000 charging-discharging cycles while maintaining the specific capacitance as well as the current response within a given potential window. However, the introduction of pseudocapacitive materials often leads to material degradation over continuous charging-discharging (less than 10,000 cycles) and hence suffers from capacitance loss.

5.5 ELECTRODE MATERIALS FOR SUPERCAPACITORS

Nanostructured carbon (porous particles powder, fibers, fabrics, carbon nanotubes, graphene derivatives and other structures), metal oxides and conducting polymers have been known as electrode materials in SC cells for more than one decade now. Meanwhile different composites based on these materials have been investigated recently.

5.5.1 CARBON-BASED MATERIALS

In this chapter, we are mainly concerned with SCs based on various carbon materials. Carbon materials are chosen as the active electrode materials in EDLCs because of their high electrical conductivity, high specific surface area (SSA), tunable porosity and relatively low production cost. Carbon materials with high surface areas, such as the activated carbons and porous carbons, biomass-derived nanocarbons, template carbons, carbide-derived carbons (CDCs), carbon onions, single-walled carbon nanotubes (SWCNTs), multi-walled carbon nanotubes (MWCNTs), carbon nanofibers and graphene are currently hot research topics for storage applications.

5.5.1.1 Activated Carbons

Among the carbon materials, activated carbons (ACs) are currently the most used active materials in commercialised EDLCs. They make for potential electrode materials because of their unmatched electrical conductivity and very high surface area. They are easy to access, and their synthesis procedures are well established. ACs are generated by either physical, chemical or a combination of both activations of a wide variety of carbonaceous materials (e.g. wood, coal, nutshell, etc.), which helps them in achieving large accessible surface areas. Physical activation mainly uses heat treatment (from 700°C to 1200°C) in oxidising gases to produce porous structures [7]. Chemical activations are conducted at lower temperatures (from 400°C to 700°C) by making use of some strong acids (e.g. H_3PO_4) or alkalines (e.g. KOH), or corrosive salts (e.g. $ZnCl_2$) [8–10]. Depending on the precursor choice and activation procedure used, the Brunauer-Emmet-Teller(BET) surface areas of the ACs can vary from 500 to 3000 $m^2\ g^{-1}$ and have different levels of porosity [7]. According to the guidelines of the International Union of Pure and Applied Chemistry (IUPAC), there can be three different kinds of pore size distribution: micropores (<2 nm), mesopores (2–50 nm) or macropores (>50 nm). Theoretically, a higher surface area of the active electrode generally allows for more charge storage, leading to higher specific capacitance of the device. In reality, however, the relationship is not so straightforward, mainly because of the broad distribution of porosity in ACs, as not all of these pores are effective for supercapacitive energy storage. Aspects such as pore size distribution, material precursor, electrolyte ion size, surface wettability and pore accessibility also need to be considered when evaluating a potential electrode material. Excessive pores not only increase the volume of the electrode materials but also hinder the electrical conductivity. Since micropores have higher surface-to-volume ratio, they tend to produce higher surface areas, most suitable for EDLC applications. Most of the aqueous electrolyte ions could access the micropores around 1 nm, while organic electrolyte cannot, due to the larger effective size of the electrolyte ions in organic solutions. Therefore, when an AC-based supercapacitor electrode is tested in aqueous electrolytes, the specific capacitance is higher than that tested in organic electrolytes (100–300 F g^{-1} versus less than 150 F g^{-1}). However, most of the commercial SCs use organic electrolytes for higher operating voltages. Thus, the main challenge of using ACs as SC electrodes is to narrow down the pore size distribution and to eliminate micropores and macropores.

5.5.1.2 Biomass-Derived Nanocarbon

Biomass-derived nanocarbons are mostly activated and porous carbon structures especially derived from lignocellulosic biomass materials, such as wood, nutshell wastes, etc. Biowaste materials are currently the most sought after because they are abundant, need to be recycled and are ample sources of carbon. Depending on the precursor structure as well as activation technique, a number of carbonaceous structures can be obtained, for example, activated carbons, carbon onions, few-layer graphene, etc. Recently, a surge of biomass-derived carbon-based research is being directed to SC applications. A number of biowastes, for example, plant leaves [11], seaweeds [12], fungi [13], wasted coffee beans [14] and even human hair [15,16] and urine [17] have been utilised to derive activated carbons obtained via physical or chemical activation methods for SC applications. Some of the researchers even extracted few-layer graphene from biomasses, like coconut shell [18], soybean shells [19], wheat-straw [20] or tea-tree plant [21] by using graphitising agents (FeCl$_3$) or sophisticated high-temperature furnaces, like in chemical vapor deposition (CVD). Graphene is a far more superior candidate as compared to activated and porous carbons, both in terms of conductivity and specific surface area. More recently, a high surface area few-layer graphene-like structure has been obtained via a simple mechanical exfoliation technique without the need for harsh chemicals or high-end instrumentation from an abundant biowaste source, peanut shells [22]. Figure 5.5 describes the schematic synthetic procedure highlighting the unique exfoliated technique, that resulted in a high-quality few-layer graphene with remarkable surface area. The precursor internal structure as well as fabrication technique often leads to the desired morphology with a controlled ratio of microporosity-to-mesoporosity to ensure the maximum application potential of biomass-derived nanocarbons for energy storage applications.

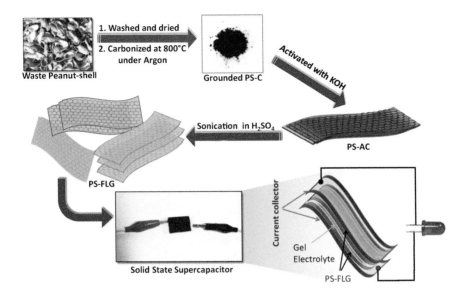

FIGURE 5.5 Synthesis of PS-FLG active material and its subsequent integration into a solid-state device. (Reproduced from T. Purkait et al., *Scientific Reports* 7(1), 2017, 1–14.)

5.5.1.3 Graphene

Graphene has attracted overwhelming research attention since its first isolation from natural graphite [23]. It is an extremely promising two-dimensional (2D) nanomaterial of an atomically thin layer of carbon with excellent mechanical and electrical properties. It has inherent structural flexibility yet substantial mechanical strength, optical transparency and enormous theoretical surface area (2630 m^2 g^{-1}) [24,25], which is why it is an excellent choice to be studied as electrode material for SCs [25,26]. The main challenge for adapting graphene material for SC application is to overcome the restacking problem. Compared to physically grown graphene (e.g. CVD growth), chemically derived graphene sheets tend to stack (due to van der Waals interactions between the layers) and suffer from overall low electrical conductivity [27,28].

Electrochemical reduction of graphene oxide (GO) to 3D graphene or, more precisely, reduced graphene oxide (3DrGO) is one of the most convenient techniques to prevent restacking and produce a porous 3D structure suitable for SC applications. Recently, a number of researchers have developed 3D graphene from GO via electrochemical cathodic cycling [29], constant potential reduction [30,31] or electrophoretic deposition [32,33]. Dey et al. reported 3DrGO from GO by using an electrochemical method of repetitive cathodic potential cycling on a copper foam template, which exhibited a specific capacitance of 623 ± 7 F g^{-1} at a current density of 1 A g^{-1} in a three-electrode system with aqueous electrolyte [29].

Another recently investigated method to prevent restacking of graphene or reduced graphene oxide (rGO) sheets is to place spacers among the individual sheets. The spacers should be electrically conductive, porous and electrochemically active. The well-studied carbon materials as spacers include carbon black particles, templated carbon spheres, CNTs and ACs [34]. For example, a carbon black–rGO hybrid with uniform structure could be obtained from a solution-based self-assembly, with the conductive carbon black particles as spacers [35]. This method makes use of the electrostatic attraction between the carbon black particles and the GO sheets that have opposite surface charges to drive the self-assembly process. After chemical reduction, the hybrid structure with much less restacking has shown 70% improvement in specific capacitance as compared to the rGO electrode without carbon black [27].

5.5.1.4 Other Carbonaceous Structures

Some other carbonaceous structures such as carbon nanofibers, carbon-nanotubes (CNTs) or onion-like carbon (OLC) have also been studied for their storage capacity. OLCs are obtained by the graphitisation of nanodiamonds (heat treatment at 1500°C). The achievable capacities remain low, as the surface is not highly developed (maximum surface area 500 m^2 g^{-1}), but these materials can deliver large power due to the full access to the carbon surface, since these nanoparticles (5 nm diameter) are non-porous [36] CNTs are also interesting to study as they have unique properties arising from their nanotubular structure and excellent electrical conductivity. However, the capacitance values are typically very low (20–80 F g^{-1}) [37]. In order to improve the performance of CNTs, additional activation or surface functionalization can be undertaken, and capacitance values can be improved to ∼80–130 F g^{-1} [38].

5.5.2 Hybrid Materials

5.5.2.1 Nanocarbon-Conducting Polymer Composites

Electrically conducting polymers (ECPs) are one of the most commonly studied pseudocapacitive material for SCs. The charge storage mechanism in ECP materials is based on the creation of structural defects resulting from the oxidation (or reduction) of the polymer chain. The terms *p-type* or *n-type doping* refer to the oxidation or reduction of a neutral polymer, respectively. A number of conducting polymers, such as polyaniline, polypyrrole or polythiophene, have been on the research front for quite some time now as pseudocapacitive storage materials [39–43]. But, the electrochemical performance of conducting polymers is quite limited in terms of voltage, energy density and power. Along with these disadvantages, the number of charge-discharge cycles is limited (<10,000) due to the mechanical degradation of the electrode by loss of material.

To improve their performance, they are often hybridised with carbon structures. The synergistic effect of capacitative property of the carbon skeleton and the faradaic contribution of the ECPs improve the overall specific capacitance, energy density as well as cycling stability. There has been a marked increase in the number of research studies in this field over the last 3 years. Recent reports suggest remarkable performances with ECP-carbon hybrid structures, such as PANI-GO [43], PPy-rGO [43], PPy-GF [39] etc. PANI is the most studied ECP for SC applications with a broad range of specific capacitance (30–3000 F/g), mainly due to polymer structure, dopant level, morphologies and ionic diffusion caused by the polymerisation process [44,45].

Similarly, PPy is an additionally known ECP for its flexibility and ease of synthesis. Due to its higher packing densities, it shows lower gravimetric capacitance but has very high potential volumetric capacitance [46]. They are often used in a composite structure with graphene or CNT. Recent results show exceedingly good storage performances of these composite structures with unprecedented volumetric parameters [42,47–49].

5.5.2.2 Nanocarbon–Metal Oxide Hybrids

Metal oxides are still the most studied pseudocapacitive materials and probably the materials with the greatest potential for industrial development. For this reason, they must meet certain specifications: a simple synthesis process from low-cost materials and good power and energy densities (because of participating Faradaic reactions).

Starting from ruthenium oxide (RuO_2) as a SC electrode in aqueous H_2SO_4 electrolyte in 1971, many transition metal oxides have been explored. An SC device based on a $RuO_2.xH_2O$ exhibited a specific capacitance of 995 F g^{-1} at 1 mV s^{-1} scan rate [50]. Manganese dioxide, MnO_2, is, in this respect, another pseudocapacitive material of choice. It is naturally abundant, inexpensive and not too toxic [51]. Its charge storage mechanism has been described as involving very fast Mn^{3+}/Mn^{4+} redox reactions with cation exchanges, protons and cations of the electrolyte, to balance the change in the oxidation state of manganese [52]:

$$MnO_2 + (x + y)\,e^- + xH^+ + yC^+ \Rightarrow MnOOH_xC_y$$

where C^+ is the cation from the electrolyte.

The literature reports capacitances of approximately 120–250 F/g, and cycling has been demonstrated over several hundred thousand charge/discharge cycles, usually lasting several dozen seconds [53]. The energy density is greater than 5 W h kg^{-1}. It operates in aqueous electrolytes (up to temperatures of $-20°C$) and is therefore safer and green.

However, the power performances of pseudocapacitive oxides are limited by their low electronic conductivity. Also, the energy of the system is then sometimes limited by the stability of the electrolyte used, usually over a potential window of 1.5–2 V maximum.

Therefore, like other pseudocapacitive materials, metal-oxides are also coupled with nanocarbons to gain from the synergistic effect. Decorating large specific surface area carbons with pseudocapacitive nano-oxides is a strategy that leads to an increase in the power density. The aim of this approach is to maximise the electronic percolation and the developed active surface area. Carbonaceous 2D (like graphene) and 1D (CNTs or carbon fibers) materials offer many opportunities for designing/shaping electrodes. Numerous publications report 'remarkable' performances, especially in terms of power, with metal oxides and hydroxides on graphene, such as, for example, with MnO_2 [54], Co_3O_4 [55] or $Ni(OH)_2$ [56].

5.6 APPLICATIONS OF SUPERCAPACITORS

SCs have mainly two kinds of applications depending on the requirement of capacitance. In small formats (with capacitance less than 50 F), SCs have been widely used for over 20 years to power small-scale electronics, for example, in small tools or in toys or to supply energy to sensors. The real turning point came with their use as the power supply for emergency systems in the opening of the doors of the Airbus A380, a programme that began in 2005 and was developed from the design of the aircraft.

Since then, the SC market has continued to develop both in small formats as well as in those with greater capacitances (several thousand Farads). Today they are being used in network regulation in tramways or trolleybus lines, or to boost charging with brake energy recovery in the form of modules of several tens or hundreds of volts.

SCs have also emerged in the automobile transport market. Maxwell SCs have been used in a 1.2 kW starter-alternator supplied by Continental for Citroën (a French automobile manufacturer) since 2011. The starter-alternator helps to perform stop/start functions and regenerative braking.

In summary, SC technology has now matured, and many manufacturers explore the world of SC cells and modules. Although a number of companies are developing SCs, the principal competitors are Maxwell Technologies, Panasonic, NessCap, LS Mtron, and le groupe Bolloré.

5.7 FUTURE PERSPECTIVE: RECENT ADVANCES AND DISADVANTAGES

5.7.1 ADVANCES OF SUPERCAPACITORS

SCs have some of the most remarkable advantages when compared to other energy storage devices, such as lithium ion batteries, including the following.

1. *High Power Density and Fast Charging*: As mentioned earlier, SCs have much higher power density than that of batteries. In case of EDLCs, the electrical charges are stored mainly at the electrode-electrolyte interface [2]. Therefore, the charge/discharge process will not be limited by the ionic conduction into the bulk of the electrode. This results in much higher charge and discharge rates (e.g. a few seconds for supercapacitors versus a few hours for batteries.). As shown in Equation 5.4, the power density is inversely proportional to the discharge time, therefore, the rapid discharge rates of SCs lead to high power density [2]. Therefore, SCs have been finding their applications in systems where rapid charge/discharge is needed, such as in energy recovery, regenerative braking or burst-mode power delivery systems within trains, automobiles, cranes and elevators [57].
2. *Excellent Cycling Ability and Shelf Life*: Due to the unique energy storage mechanisms of SCs, the charge-discharge process is highly reversible. Whether it is the physisorption of electrolyte ions onto a porous electrode for EDLC, or the faradic reactions for pseudocapacitors, there is no chemical bond breaking involved. Therefore, typical SCs could be cycled at high rates for 10,000–1,000,000 cycles with minimal changes of their electrochemical properties [58]. In contrast, the energy storage in a typical battery often involves irreversible intercalations and phase changes of the electrode materials. Thus, the cycle life for batteries is incomparable to that of SCs, even when the stored energy is as small as 10%–20% of the overall potential.

Supercapacitor

Another advantage of supercapacitors related to their high chemical stability is the long shelf life. After long-term storage (a few months to a few years), supercapacitors can still maintain their capacitance and recharging ability. On the other hand, most rechargeable batteries will degrade over time when not in use, due to self-discharge and corrosion problems [59].

5.7.2 Disadvantages of Supercapacitors

Despite the above-mentioned advantages, supercapacitors still face several challenges at the current stage. These challenges could not be overlooked, if they are to find their place in the rapidly expanding energy storage market.

1. *Low Energy Density*: One of the major drawbacks of supercapacitor is the insufficient energy density, generally, in the range of 0.1–10 W h kg^{-1} [60–62], which is one to two orders of magnitude below that of the commercialized lithium-ion batteries (50–200 W h kg^{-1}) [63]. This means, to provide the same amount of energy, a much bigger or heavier supercapacitor is required, in order to replace a lithium-ion battery. To overcome this challenge, considerable research work has been focused on improving the specific capacitance of electrode materials in the past few years. Another approach is to increase the voltage window through fabrication of asymmetric supercapacitor.
2. *High Cost*: The cost of electrode materials for supercapacitors represent a significant share of the total cost. This is another challenge for the scaling up and commercialization of supercapacitors. Currently, the most common commercially available electrode materials are mostly activated carbons. The cost of carbon nanomaterials with high surface area that could be used for EDLC has dropped from of US $50–100 per kg to US $15 per kg recently, based on a study in 2013 [64]. However, due to the low energy density, the overall energy storage cost by supercapacitors is US $2400–6000 per kW h while the one for lithium ion battery is US $500–1000 per kW h, based on the well-to-wheel analysis in 2011.

In addition to the electrode materials, some electrolytes, such as organic or ionic liquid electrolytes, are also of high cost, although they can help to extend the operation voltage window. Therefore, they also contribute to the overall cost of energy storage.

REFERENCES

1. B. Howard, Low voltage electrolytic capacitor, U.S., Pat., 1957, 2800616. http://www.freepatentsonline.com/2800616.html%0Ahttps://worldwide.espacenet.com/publicationDetails/biblio?CC=US&NR=2800616&KC=&FT=E&locale=en_EP.
2. G. Wang, L. Zhang, J. Zhang, A review of electrode materials for electrochemical supercapacitors, *Chem. Soc. Rev.* 41, 2012, 797–828. doi: 10.1039/c1cs15060j
3. H. Helmholtz, Der phpsik und chemie., *Ann. Der Phys. Und Chemie.* 7, 1879, 22.
4. O. Stern, Zur Theorie der Elektrolytischen Doppelschicht, *Z. Electrochem.* 30, 1924, 508–516. doi: 10.1002/bbpc.192400182

5. B.E. Conway, Transition from "Supercapacitor" to "Battery" behavior in electrochemical energy storage, *J. Electrochem. Soc.* 138, 1991, 1539. doi: 10.1149/1.2085829
6. Y.Z. Zhang, Y. Wang, T. Cheng, W.Y. Lai, H. Pang, W. Huang, Flexible supercapacitors based on paper substrates: A new paradigm for low-cost energy storage, *Chem. Soc. Rev.* 44, 2015, 5181–5199. doi: 10.1039/c5cs00174a
7. L.L. Zhang, X.S. Zhao, Carbon-based materials as supercapacitor electrodes, *Chem. Soc. Rev.* 38, 2009, 2520–2531. doi: 10.1039/b813846j
8. K. Kierzek, E. Frackowiak, G. Lota, G. Gryglewicz, J. Machnikowski, Electrochemical capacitors based on highly porous carbons prepared by KOH activation, *Electrochim. Acta.* 49, 2004, 515–523. doi: 10.1016/j.electacta.2003.08.026
9. O. Barbieri, M. Hahn, A. Herzog, R. Kötz, Capacitance limits of high surface area activated carbons for double layer capacitors, *Carbon N. Y.* 43, 2005, 1303–1310. doi: 10.1016/j.carbon.2005.01.001
10. E. Raymundo-Piñero, K. Kierzek, J. Machnikowski, F. Béguin, Relationship between the nanoporous texture of activated carbons and their capacitance properties in different electrolytes, *Carbon N. Y.* 44, 2006, 2498–2507. doi: 10.1016/j.carbon.2006.05.022
11. M. Biswal, A. Banerjee, M. Deo, S. Ogale, From dead leaves to high energy density supercapacitors, *Energy Environ. Sci.* 6, 2013, 1249–1259. doi: 10.1039/c3ee22325f
12. E. Raymundo-Piñero, M. Cadek, F. Béguin, Tuning carbon materials for supercapacitors by direct pyrolysis of seaweeds, *Adv. Funct. Mater.* 19, 2009, 1032–1039. doi: 10.1002/adfm.200801057
13. H. Zhu, X. Wang, F. Yang, X. Yang, Promising carbons for supercapacitors derived from fungi, *Adv. Mater.* 23, 2011, 2745–2748. doi: 10.1002/adma.201100901
14. T.E. Rufford, D. Hulicova-Jurcakova, Z. Zhu, G.Q. Lu, Nanoporous carbon electrode from waste coffee beans for high performance supercapacitors, *Electrochem. Commun.* 10, 2008, 1594–1597. doi: 10.1016/j.elecom.2008.08.022
15. R. Satish, A. Vanchiappan, C.L. Wong, K.W. Ng, M. Srinivasan, Macroporous carbon from human hair: A journey towards the fabrication of high energy Li-ion capacitors, *Electrochim. Acta.* 182, 2015, 474–481. doi: 10.1016/j.electacta.2015.09.127
16. W. Qian, F. Sun, Y. Xu, L. Qiu, C. Liu, S. Wang, F. Yan, Human hair-derived carbon flakes for electrochemical supercapacitors, *Energy Environ. Sci.* 7, 2013, 379–386. doi: 10.1039/C3EE43111H
17. N.K. Chaudhari, M.Y. Song, J.-S. Yu, Heteroatom-doped highly porous carbon from human urine, *Sci. Rep.* 4, 2014, 5221. doi: 10.1038/srep05221
18. L. Sun, C. Tian, M. Li, X. Meng, L. Wang, R. Wang, J. Yin, H. Fu, From coconut shell to porous graphene-like nanosheets for high-power supercapacitors, *J. Mater. Chem. A.* 1, 2013, 6462–6470. doi: 10.1039/c3ta10897j
19. H. Zhou, J. Zhang, I.S. Amiinu, C. Zhang, X. Liu, W. Tu, M. Pan, S. Mu, Transforming waste biomass with an intrinsically porous network structure into porous nitrogen-doped graphene for highly efficient oxygen reduction, *Phys. Chem. Chem. Phys.* 18, 2016, 10392–10399. doi: 10.1039/C6CP00174B
20. F. Chen, J. Yang, T. Bai, B. Long, X. Zhou, Facile synthesis of few-layer graphene from biomass waste and its application in lithium ion batteries, *J. Electroanal. Chem.* 768, 2016, 18–26. doi: 10.1016/j.jelechem.2016.02.035
21. M.V. Jacob, R.S. Rawat, B. Ouyang, K. Bazaka, D.S. Kumar, D. Taguchi, M. Iwamoto, R. Neupane, O.K. Varghese, Catalyst-free plasma enhanced growth of graphene from sustainable sources, *Nano Lett.* 15, 2015, 5702–5708. doi: 10.1021/acs.nanolett.5b01363
22. T. Purkait, G. Singh, M. Singh, D. Kumar, R.S. Dey, Large area few-layer graphene with scalable preparation from waste biomass for high-performance supercapacitor, *Sci. Rep.* 7, 2017, 1–14. doi: 10.1038/s41598-017-15463-w
23. A.K. Geim, K.S. Novoselov, The rise of graphene, *Nat. Mater.* 6, 2007, 183–191. doi: 10.1038/nmat1849

24. J. Avila, I. Razado, S. Lorcy, R. Fleurier, E. Pichonat, D. Vignaud, X. Wallart, M.C. Asensio, Exploring electronic structure of one-atom thick polycrystalline graphene films: A nano angle resolved photoemission study, *Sci. Rep.* 3, 2013, 1–8. doi: 10.1038/srep02439
25. M. Pumera, Graphene-based nanomaterials for energy storage, *Energy Environ. Sci.* 4, 2011, 668–674. doi: 10.1039/c0ee00295j
26. F. Bonaccorso, L. Colombo, G. Yu, M. Stoller, V. Tozzini, A.C. Ferrari, R.S. Ruoff, V. Pellegrini, Graphene, related two-dimensional crystals, and hybrid systems for energy conversion and storage, *Science* 347, 2015, 1246501. doi: 10.1126/science.1246501
27. Z. Lei, J. Zhang, L.L. Zhang, N.A. Kumar, X.S. Zhao, Functionalization of chemically derived graphene for improving its electrocapacitive energy storage properties, *Energy Environ. Sci.* 9, 2016, 1891–1930. doi: 10.1039/c6ee00158k
28. M.D. Stoller, S. Park, Y. Zhu, J. An, R.S. Ruoff, Graphene-based ultracapacitors, *Nano Lett.* 8, 2008, 3498–3502. doi: 10.1021/nl802558y
29. R.S. Dey, H.A. Hjuler, Q. Chi, Approaching the theoretical capacitance of graphene through copper foam integrated three-dimensional graphene networks, *J. Mater. Chem. A.* 3, 2015, 6324–6329. doi: 10.1039/C5TA01112D
30. Y. Li, K. Sheng, W. Yuan, G. Shi, A high-performance flexible fibre-shaped electrochemical capacitor based on electrochemically reduced graphene oxide, *Chem. Commun.* 49, 2013, 291–293. doi: 10.1039/C2CC37396C
31. T. Purkait, G. Singh, D. Kumar, M. Singh, R.S. Dey, High-performance flexible supercapacitors based on electrochemically tailored three-dimensional reduced graphene oxide networks, *Sci. Rep.* 8, 2018, 640. doi: 10.1038/s41598-017-18593-3
32. M. Wang, L.D. Duong, J.S. Oh, N.T. Mai, S. Kim, S. Hong, T. Hwang, Y. Lee, J. Do Nam, Large-area, conductive and flexible reduced graphene oxide (RGO) membrane fabricated by electrophoretic deposition (EPD), *ACS Appl. Mater. Interfaces.* 6, 2014, 1747–1753. doi: 10.1021/am404719u
33. K. Shi, I. Zhitomirsky, Electrophoretic nanotechnology of graphene-carbon nanotube and graphene-polypyrrole nanofiber composites for electrochemical supercapacitors, *J. Colloid Interface Sci.* 407, 2013, 474–481. doi: 10.1016/j.jcis.2013.06.058
34. A. Ghosh, Y.H. Lee, Carbon-based electrochemical capacitors, *ChemSusChem.* 5, 2012, 480–499. doi: 10.1002/cssc.201100645
35. C.X. Guo, C.M. Li, A self-assembled hierarchical nanostructure comprising carbon spheres and graphene nanosheets for enhanced supercapacitor performance, *Energy Environ. Sci.* 4, 2011, 4504–4507. doi: 10.1039/c1ee01676h
36. P. Simon, Y. Gogotsi, Capacitive energy storage in nanostructured carbon-electrolyte systems, *Acc. Chem. Res.* 46, 2013, 1094–1103. doi: 10.1021/ar200306b
37. S. Talapatra, S. Kar, S.K. Pal, R. Vajti, L. Ci, P. Victor, M.M. Shaijumon, S. Kaur, O. Nalamasu, P.M. Ajayan, Direct growth of aligned carbon nanotubes on bulk metals, *Nat. Nanotechnol.* 1, 2006, 112–116. doi: 10.1038/nnano.2006.56
38. C. Emmenegger, P. Mauron, P. Sudan, P. Wenger, V. Hermann, R. Gallay, A. Züttel, Investigation of electrochemical double-layer (ECDL) capacitors electrodes based on carbon nanotubes and activated carbon materials, *J. Power Sources.* 124, 2003, 321–329. doi: 10.1016/S0378-7753(03)00590-1
39. Y. Lee, H. Choi, M.S. Kim, S. Noh, K.J. Ahn, K. Im, O.S. Kwon, H. Yoon, Nanoparticle-mediated physical exfoliation of aqueous-phase graphene for fabrication of three-dimensionally structured hybrid electrodes, *Sci. Rep.* 6, 2016, 1–10. doi: 10.1038/srep19761
40. H. Hu, K. Zhang, S. Li, S. Ji, C. Ye, Flexible, in-plane, and all-solid-state micro-supercapacitors based on printed interdigital Au/polyaniline network hybrid electrodes on a chip, *J. Mater. Chem. A.* 2, 2014, 20916–20922. doi: 10.1039/c4ta05345a
41. P. Li, Y. Yang, E. Shi, Q. Shen, Y. Shang, S. Wu, J. Wei, et al. Core-double-shell, carbon nanotube@polypyrrole@MnO$_2$ sponge as freestanding, compressible supercapacitor electrode, *ACS Appl. Mater. Interfaces.* 6, 2014, 5228–5234. doi: 10.1021/am500579c

42. L. Yuan, B. Yao, B. Hu, K. Huo, W. Chen, J. Zhou, Polypyrrole-coated paper for flexible solid-state energy storage, *Energy Environ. Sci.* 6, 2013, 470–476. doi: 10.1039/c2ee23977a
43. Z.S. Wu, K. Parvez, S. Li, S. Yang, Z. Liu, S. Liu, X. Feng, K. Müllen, Alternating stacked graphene-conducting polymer compact films with ultrahigh areal and volumetric capacitances for high-energy micro-supercapacitors, *Adv. Mater.* 27, 2015, 4054–4061. doi: 10.1002/adma.201501643
44. K. Naoi, S. Suematsu, Y. Oura, H. Tsujimoto, H. Kanno, Conducting polymer films of cross-linked structure and their QCM analysis, *Electrochim. Acta.* 45, 2000, 3813–3821.
45. E. Frackowiak, V. Khomenko, K. Jurewicz, K. Lota, F. Béguin, Supercapacitors based on conducting polymers/nanotubes composites, *J. Power Sources.* 153, 2006, 413–418. doi: 10.1016/j.jpowsour.2005.05.030
46. Y. Shi, L. Pan, B. Liu, Y. Wang, Y. Cui, Z. Bao, G. Yu, Nanostructured conductive polypyrrole hydrogels as high-performance, flexible supercapacitor electrodes, *J. Mater. Chem. A.* 2, 2014, 6086–6091. doi: 10.1039/c4ta00484a
47. Z.S. Wu, Y. Zheng, S. Zheng, S. Wang, C. Sun, K. Parvez, T. Ikeda, X. Bao, K. Müllen, X. Feng, Stacked-layer heterostructure films of 2d thiophene nanosheets and graphene for high-rate all-solid-state pseudocapacitors with enhanced volumetric capacitance, *Adv. Mater.* 29, 2017, 1–7. doi: 10.1002/adma.201602960
48. M. Beidaghi, C. Wang, Micro-supercapacitors based on three dimensional interdigital polypyrrole/C-MEMS electrodes, *Electrochim. Acta.* 56, 2011, 9508–9514. doi: 10.1016/j.electacta.2011.08.054
49. C.J. Raj, B.C. Kim, W.J. Cho, W.G. Lee, S.D. Jung, Y.H. Kim, S.Y. Park, K.H. Yu, Highly flexible and planar supercapacitors using graphite flakes/polypyrrole in polymer lapping film, *ACS Appl. Mater. Interfaces.* 7, 2015, 13405–13414. doi: 10.1021/acsami.5b02070
50. M.Y. Ho, P.S. Khiew, D. Isa, T.K. Tan, A. Materials, J. Broga, J.G. Kelang et al. Science, a review of metal oxide composite electrode materials for, NANO Br., *Reports Rev.* 9, 2014, 1–25. doi: 10.1142/S1793292014300023
51. T. Brousse, B. Daniel, To be or not to be pseudocapacitive? *J. Electrochem. Soc.*, 162, 2015, 5185–5189. doi: 10.1149/2.0201505jes
52. M. Toupin, T. Brousse, D. Bélanger, Charge storage mechanism of MnO_2 electrode used in aqueous electrochemical capacitor, *Chem. Mater.*, 16, 2004, 3184–3190. doi: 10.1021/cm049649j
53. T. Brousse, P.L. Taberna, O. Crosnier, R. Dugas, P. Guillemet, Y. Scudeller, Y. Zhou, F. Favier, D. Bélanger, P. Simon, Long-term cycling behavior of asymmetric activated carbon/MnO_2 aqueous electrochemical supercapacitor, *J. Power Sources.* 173, 2007, 633–641. doi: 10.1016/j.jpowsour.2007.04.074
54. S. Chen, J. Zhu, X. Wu, Q. Han, X. Wang, Graphene oxide MnO_2, *ACS Nano.* 4, 2010, 2822–2830. doi: 10.1021/nn901311t
55. J. Yan, T. Wei, W. Qiao, B. Shao, Q. Zhao, L. Zhang, Z. Fan, Rapid microwave-assisted synthesis of graphene nanosheet/Co_3O_4 composite for supercapacitors, *Electrochim. Acta.* 55, 2010, 6973–6978. doi: 10.1016/j.electacta.2010.06.081
56. H. Wang, H.S. Casalongue, Y. Liang, H. Dai, $Ni(OH)_2$ nanoplates grown on graphene as advanced electrochemical pseudocapacitor Materials, *J. Am. Chem. Soc.* 132, 2010, 7472–7477. doi: 10.1021/ja102267j
57. D. Kalpana, T. Thyagarajan, R. Thenral, Improved identification and control of 2-by-2 MIMO system using relay feedback, *Control Eng. Appl. Informatics.* 17, 2015, 23–32. doi: 10.1016/j.isatra.2015.09.012
58. M. Inagaki, H. Konno, O. Tanaike, Carbon materials for electrochemical capacitors, *J. Power Sources.* 195, 2010, 7880–7903. doi: 10.1016/j.jpowsour.2010.06.036
59. A. Burke, Ultracapacitors: Why, how, and where is the technology, *J. Power Sources.* 91, 2000, 37–50. doi: 10.1016/S0378-7753(00)00485-7

60. G. Qu, J. Cheng, X. Li, D. Yuan, P. Chen, X. Chen, B. Wang, H. Peng, A fiber supercapacitor with high energy density based on hollow graphene/conducting polymer fiber electrode, *Adv. Mater.* 28, 2016, 3646–3652. doi: 10.1002/adma.201600689
61. M. Yu, Z. Wang, Y. Han, Y. Tong, X. Lu, S. Yang, Recent progress in the development of anodes for asymmetric supercapacitors, *J. Mater. Chem. A.* 4, 2016, 4634–4658. doi: 10.1039/c5ta10542k
62. P. Simon, Y. Gogotsi, Materials for electrochemical capacitors, *Nat. Mater.* 7, 2008, 845–854. doi: 10.1038/nmat2297
63. L. Damen, J. Hassoun, M. Mastragostino, B. Scrosati, Solid-state, rechargeable Li/LiFePO4 polymer battery for electric vehicle application, *J. Power Sources.* 195, 2010, 6902–6904. doi: 10.1016/j.jpowsour.2010.03.089
64. W.K. Chee, H.N. Lim, Z. Zainal, N.M. Huang, I. Harrison, Y. Andou, Flexible graphene-based supercapacitors: A review, *J. Phys. Chem. C.* 120, 2016, 4153–4172. doi: 10.1021/acs.jpcc.5b10187
65. J.J. Yoo, K. Balakrishnan, J. Huang, V. Meunier, B.G. Sumpter, A. Srivastava, M. Conway et al., Ultrathin planar graphene supercapacitors, *Nano Lett.* 4, 2011, 1423–1427.

6 Solar Cell

6.1 INTRODUCTION

Solar cells are such devices where semiconductor materials produce electricity in the presence of sunlight; the phenomenon is known as the photovoltaic effect. The fuel of solar cells is sunlight, which is the most abundant element in the globe and it demands no cost. When sunlight falls into a semiconductor, it produces a pair of negatively charged electrons and a positively charged hole. These charged particles disperse throughout the semiconductor and eventually encounter an energy barrier that allows charged particles of one sign to penetrate but reflect those of the other sign. As a result, a bunch of electrons flow through the metal to the electric load and produce electricity. Therefore, solar cell technologies that can consume solar light and produce electricity are of utmost interest in the field of renewable energies [1]. Various nanomaterials including nanocarbons have potential for the development of solar cells.

It is considered that more than 90% of the solar cell market is dependent on crystalline silicon (both single-crystal and polycrystalline silicon), which makes it a costly material all over the world [2]. Considering the present climate change issues, it is therefore necessary to search for some new and cost-effective renewable energy materials. Carbon-based solar cells have gained considerable interest in the last two decades in both fundamental science as well as from technological points of view due to their potential application [3]. Nanocarbon exists in the form of several allotropes including graphene, carbon nanotubes (CNTs), activated carbon, etc. with unique optical, electrical, optoelectronics and mechanical properties, which makes them suitable for use in solar cells [4]. Graphene and carbon nanotubes and their composites were appraised as highly competitive materials to replace the transparent conductive oxide plate. Furthermore, carbon nanomaterials, specifically graphene, have proven to be high electron transport as well as light absorption materials that can be used with semiconducting nanostructured film in solar cells.

Different carbon nanomaterials and their composites have been used recently. Graphene and carbon nanotubes were recently been employed as counter-electrodes in making solar devices. Graphene has been proven as electrode material, which can act as an electrode and replace metal to make a solar cell free of a hole transfer layer (HTL). The role of the nanocarbon in various types of solar cell is highlighted in the following section.

6.2 DIFFERENT TYPES OF PHOTOVOLTAIC DEVICES

There are different kinds of solar cells based on the materials used to make the device. All the different kinds of materials used in the solar cell must have evident

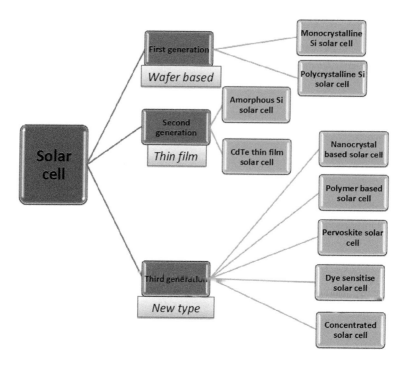

FIGURE 6.1 Different types of photovoltaic technologies developed in recent years.

characteristics so as to absorb the light and produce electricity. Earlier solar cells are made with thin silicon wafer. The current photovoltaics are categorised based on their electron hole creation mechanism and type of semiconducting material used in the solar cell [5]. The following steps occur in all types of solar cells. First, they absorb sunlight, and separation of charges (electrons and holes) takes place. As a result, generation of electric current and voltage happens and the cell works like a battery under sunlight. Different types of photovoltaic technologies developed over the years are shown in Figure 6.1. A few of the different types of photovoltaic devices and the role of carbon nanomaterials in the cell are discussed in the following section.

6.3 SILICON-BASED SOLAR CELLS

Silicon has a bandgap of 1.12 eV, and the light absorption cut-off wavelength is about 1160 nm. This bandgap of silicon is fortunately matched with the solar spectrum and is appropriate to the optimal value for solar-to-electric energy conversion with a single semiconductor optical absorber [5]. Silicon solar cells are the most-used solar cell technology and are dominating the global market. They are expected to control the photovoltaic market in the near future [6]. Silicon solar cells are considered to have a high power conversion efficiency (PCE) of about 15%, but the cost of making the solar cell is not inexpensive due to the limited sources of silicon [7]. It is therefore essential to look for some cost-effective material that can bring down the solar cell cost as well as increase the efficiency of the solar cell. Carbon and silicon are both elements that are plentiful in nature. The combination of conventional silicon semiconductors with carbon nanomaterials in the fabrication of solar cells has been a promising research field. The PCE increased from 15% to 17% with an overwhelming speed, and the variety of systems motivates the researchers in this field.

Nano-structuring of silicon has been explored by different methods such as alloying, doping and defect site engineering of the silicon surface for quantum confinement in nanostructures to improve the efficiency of the solar cell. The crystalline silicon and silicon nanowire (SiNW) have been widely investigated because of their exciting physical, optoelectronic and photovoltaic properties [8]. There are two different approaches for the growth of SiNWs – bottom-up and top-down approaches, as described in Figure 6.2. The bottom-up approach can grow the nanowire of diameters ranging from 5 to 100 nm with length 100 nm to several micrometers. However, the top-down approach for the synthesis of nanowire involves lithography and etching methods.

Various allotropes of carbon nanomaterials exist, such as carbon nanotubes (CNTs), graphene, fullerene (C_{60}), carbon nanofibers and so on. Carbon nanomaterials have several advantages such as mechanical flexibility, chemical stability, carrier mobility and huge surface area. The extraordinary properties of carbon nanomaterials meet the requirements of heterojunction-based solar cells [6]. Furthermore, silicon has a bandgap of 1.1 eV, the combination of C/Si expected to form an efficient heterostructure for photovoltaic applications. Graphene/Si heterojunction solar cells are flourishing and increased the efficiency up to 15%–17% in recent decades. The mechanism is quite simple in the case of the carbon–Si heterostructure, where a diffusion process of charge transporter occurs due to the concentration differences of holes and electron while p-type carbon and n-type silicon are hybridised together for a solar cell.

Yu et al. demonstrated the first report of a carbon/Si heterojunction solar cell with 40 nm thin carbon film deposited on n-Si by the chemical vapor deposition method [9]. The efficiency of the PCE device was 3.8%. After that Chen et al. developed C60/p-Si solar material and found that the materials have rectifying properties in the absence of light owing to the formation of a heterojunction [10]. The rectifying behaviour with a mid-infrared photocurrent response was observed with CNTs/n-Si heterojunction, and this platform opens the way to fabricate photovoltaic devices [11]. In the last decade, after the discovery of graphene, it was combined on silicon with entire coverage to form Schottky junctions, and the film was analysed for enhancement of the performance of the solar cell [12]. Several works reported with a graphene/Si heterojunction solar cell with PCE achieved up to 15.6%. The number of layers of graphene have a crucial role in modulating the work function as demonstrated by Lin et al. [13]. Heteroatom doped graphene, polymer-graphene composites and nanoparticle modified graphene have been explored and found to have potential, and the bandgap can be tuned to boost the performance of the solar cell [14–16].

6.4 POLYMER-BASED SOLAR CELLS

Polymer-based organic photovoltaic (OPV) was introduced with the aim to obtain low-cost, easy to process solar material. OPVs are mainly based on polymer

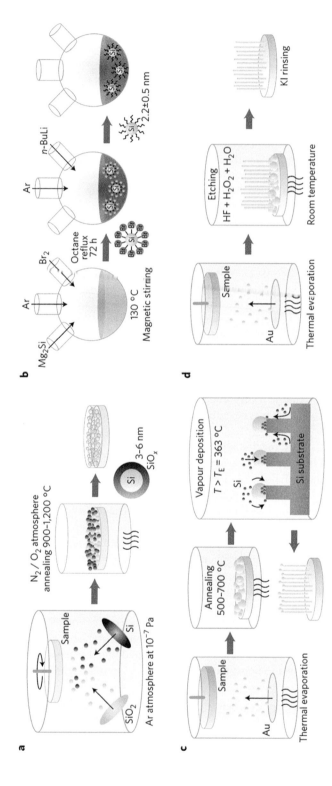

FIGURE 6.2 (a–d) The growth of SiNWs by both bottom-up and top-down approaches. (Reproduced with permission from F. Priolo et al., *Nat. Nanotechnol.* 9, 2014, 19–32. doi: 10.1038/nnano.2013.271.)

semiconductors and carbon materials. Semiconducting properties of OPV arise from electron delocalisation of the conjugated backbone due to the presence of alternating C–C and C=C bonds [17]. Engineering of the bandgap of the organic polymer material can be used for the colour change of the organic semiconductor material [18]. Recently, organic light-emitting diodes (OLEDs), organic transistors, organic memory devices and OPVs have drawn attention in the electronics market due to their use in smart and portable devices because of low cost and easy synthetic approaches [19].

Various organic semiconductor materials are used in the literature (Figure 6.3) such as MEH-PPV: poly [2-methoxy-5-(2'-ethyl-hexyloxy)-1,4-phenylene vinylene]; P3HT: poly(3-hexylthiophene); PFO-DBT: poly [2,7-(9,9-dioctyl-fluorene)-alt-5,5-(4',7'-di-2-thienyl-2',1',3'-benzothiadiazole); PCDTBT: poly[N-9'-hepta-decanyl-2,7-carbazole-alt-5,5-(4',7'-di-thienyl-2',1',3'-benzothiadizaole); CN-MEH-PPV: poly-[2-methoxy-5,2'-ethylhexyloxy]-1,4-(1-cyanovinylene)-

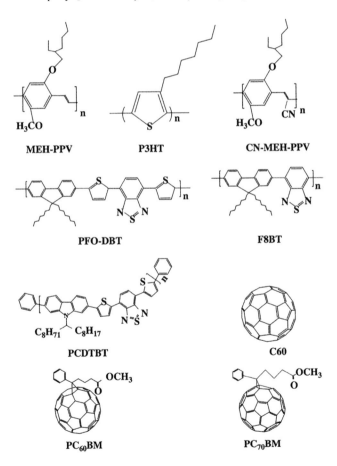

FIGURE 6.3 Example of organic semiconductors used in polymer solar cells. (Reproduced with permission from W. Cai et al., *Sol. Energy Mater. Sol. Cells*. 94, 2010, 114–127. doi: 10.1016/j.solmat.2009.10.005.)

phenylene; F8TB: poly(9,9'-dioctylfluorene-co-bis-N,N'-(4-butylphenyl)- bis-N,N'-phenyl-1,4-phenylenediamine; PC60BM: 6,6-phenyl-C61-butyric acid methyl ester; PC70BM: 6,6-phenyl-C71-butyric acid methyl ester; etc.

Stability is one of the major concerns for an organic solar cell, although it is a cost-effective material and has good efficiency. However, stability can be improved by protecting the material from oxygen and humidity as well as hybridising it with carbon nanomaterials. Fullerene, graphene and CNTs have shown potential with organic semiconductors for further improvement of device performance of OPVs, while hybridising with polymer materials [20]. Carbonaceous materials are used not only as active materials but also as electrodes in organic solar cells. Owing to its many pleasing properties such as mechanical flexibility and high conductivity, graphene is considered as a potential electrode material that can replace the indium tin oxide (ITO) or fluorine tin oxide (FTO) conducting plate, which are not flexible. CNTs and reduced graphene oxide and their composites are also well studied materials for alternative materials for organic solar cells [20–22].

6.5 DYE-SENSITISED SOLAR CELLS

A dye-sensitised solar cell (DSSC) has tremendous advantages over other solar materials. A DSSC enables the generation of electricity by absorbing photons in dye molecules. O'Regan and Grätzel in 1991 first demonstrated a DSSC based on a mesoporous TiO_2 nanoparticle. The iodide/tri-iodide (I^-/I_3^-) system is the most used redox couple due to slow recombination kinetics between electrode and titania [23,24]. ITO and FTO are the transparent conducting substrates usually used to coat the materials.

Carbon nanostructures such as graphene and CNTs have been used recently to replace the conducting substrate due to several reasons such as high conductivity, mechanical flexibility and large surface area [3]. A recently prepared transparent and conducting graphene film has transmittance of 0.94–0.96. Although the conductivity of the film does not reach the ITO plate, the robustness and simple ion diffusion make it a suitable candidate for a counter-electrode material [25]. There are many reports on carbon black, activated carbon and CNTs as counter-electrodes for DSSC. It is clear that carbon nanomaterials have a potential role as electrode materials in DSSC. Nanocarbon, mainly graphene, has traditionally been used as a conducting transparent electrode for the replacement of the counter-electrode. Research is ongoing to further develop a flexible, cost-effective and more robust electrode for DSSC.

6.6 QUANTUM DOTS–BASED SOLAR CELLS

A quantum-dot (QD) solar cell is categorised as a third-generation solar cell that has garnered significant attention in the last few years due to its virtuous optoelectronic properties [26]. The working principle of the QD solar cell is similar to the DSSC, where wide bandgap materials are sensitised with semiconductor QDs [27]. Various semiconductor QDs such as CdS, CdSe, CdTe, Cu_2S, SnS_2, PbS, Sb_2O_3, PbSe, etc. are synthesised with wide bandgap materials as sensitisers. Carbon QDs (CQDs) also possess similar optical properties as other semiconductor QDs due to broad optical bandgap originating from π-plasmon, and they are being used in solar cell applications [28]. The optical- and photo-induced redox properties of the CQDs, especially graphene QDs, play a dominating role among other CQDs arising from the presence of structural and edge defects.

6.7 PEROVSKITES SOLAR CELLS

Perovskite nomenclature has originated from materials that have similar formulas such as calcium titanate, ABX_3 (where A = organic ligands, B = Pb or Sn, X = Cl, Br, I). It is interesting to note that due to the structural phase transition of methylammonium ($CH_3NH_2^+$) ion at different temperatures, inorganic-organic perovskite received significant attention in recent years [29]. In 2009 with the first attempt of $CH_3NH_3PbX_3$ (X = Br and I) as a dye-sensitised liquid junction-type solar cell with efficiency 3%–4% [30], research in this field was triggered [31]. In order to advance the efficiency of the device, carbon nanomaterials such as fullerene, graphene and CNTs have a significant role in the perovskite-based solar cell [32].

Fullerene is a well-used nanocarbon in the perovskite solar cell as it has the capability of removing TiO_2, as it does not degrade its structure at high temperature. In parallel with fullerene, graphene quantum dot and nano graphene flakes have also shown improved efficiency and good electro-transport properties in a hybrid structure compared to the conventional perovskite material [32]. TiO_2-loaded graphene or rGO has recently been used to reduce the series resistance as well as to diminish the recombination loss in the perovskite solar cell [20]. CNT-based electrodes are more exploited than other nanocarbons in perovskite solar cells, where both multi-walled CNTs and single-walled CNTs are used as charge transport carriers and to increase the efficiency of the solar cell [33]. Recently, the graphene-based perovskite solar cell was chosen for its PCE. Graphene can act as the electron/hole transfer material as well as provide contact in the perovskite solar cell. It has also been proven that graphene can replace gold and can build a solar cell free of hole transfer material (HTL) (Figure 6.4) [34].

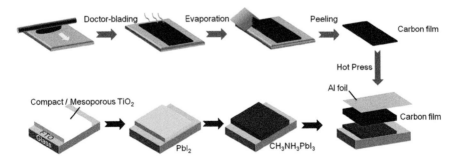

FIGURE 6.4 Carbon electrode–based HTL-free solar cell. (Reproduced with permission from H. Wei et al., *Carbon* 93, 2015, 861–868.)

6.8 FUTURE PERSPECTIVES

In this chapter, the effects of different carbon nanomaterials on the perovskite materials and their electron transport properties have been covered. It is worth mentioning that the nanocarbon not only helps to provide conductivity and mechanical strength of the perovskite materials, but it also has advantages in terms of providing enhanced stability of the solar cell. Moreover, nanocarbon reduces the recombination, which has great importance in understanding the photovoltaic effects in the perovskite solar cell. The mechanical strength of the materials can also be improved, and as a result device flexibility is easy to achieve with the carbon-based perovskite solar cell.

REFERENCES

1. P. Hersch, K. Zweibel, *Basic Photovoltaic Principles and Methods*, Solar Energy Research Institute, Golden, CO, 1982. http://www.osti.gov/servlets/purl/5191389-yrKYNd/
2. Y. Xing, P. Han, S. Wang, P. Liang, S. Lou, Y. Zhang, S. Hu, H. Zhu, C. Zhao, Y. Mi, A review of concentrator silicon solar cells, *Renew. Sustain. Energy Rev.* 51, 2015, 1697–1708.
3. L.J. Brennan, M.T. Byrne, M. Bari, Y.K. Gun'ko, Carbon nanomaterials for dye-sensitized solar cell applications: A bright future, *Adv. Energy Mater.* 1, 2011, 472–485. doi: 10.1002/aenm.201100136
4. T.H. Kim, T. Lee, W.A. El-Said, J.W. Choi, Graphene-based materials for stem cell applications, *Materials (Basel).* 8, 2015, 8674–8690. doi: 10.3390/ma8125481
5. C. Battaglia, A. Cuevas, S. De Wolf, High-efficiency crystalline silicon solar cells: Status and perspectives, *Energy Environ. Sci.* 9, 2016, 1552–1576. doi: 10.1039/c5ee03380b
6. X. Li, Z. Lv, H. Zhu, Carbon/silicon heterojunction solar cells: State of the art and prospects, *Adv. Mater.* 27, 2015, 6549–6574. doi: 10.1002/adma.201502999
7. M. Yamaguchi, K.-H. Lee, K. Araki, N. Kojima, A review of recent progress in heterogeneous silicon tandem solar cells, *J. Phys. D. Appl. Phys.* 51, 2018, 133002. doi: 10.1088/1361-6463/aaaf08
8. F. Priolo, T. Gregorkiewicz, M. Galli, T.F. Krauss, Silicon nanostructures for photonics and photovoltaics, *Nat. Nanotechnol.* 9, 2014, 19–32. doi: 10.1038/nnano.2013.271
9. H.A. Yu, Y. Kaneko, S. Yoshimura, S. Otani, Photovoltaic cell of carbonaceous film/n-type silicon, *Appl. Phys. Lett.* 547, 1995, 547. doi: 10.1063/1.116395
10. K.M. Chen, Y.Q. Jia, S.X. Jin, K. Wu, X.D. Zhang, W.B. Zhao, C.Y. Li, Z.N. Gu, The bias-temperature effect in a rectifying Nb/C60/p-Si structure: Evidence for mobile negative charges in the solid C60 film, *J. Phys. Condens. Matter.* 6, 1994. doi: 10.1088/0953-8984/6/27/002
11. M.B. Tzolov, T.F. Kuo, D.A. Straus, A. Yin, J. Xu, Carbon nanotube-silicon heterojunction arrays and infrared photocurrent responses, *J. Phys. Chem. C.* 111, 2007, 5800–5804. doi: 10.1021/jp068701r
12. X. Li, H. Zhu, K. Wang, A. Cao, J. Wei, C. Li, Y. Jia, Z. Li, X. Li, D. Wu, Graphene-on-silicon Schottky junction solar cells, *Adv. Mater.* 22, 2010, 2743–2748. doi: 10.1002/adma.200904383
13. Y. Lin, X. Li, D. Xie, T. Feng, Y. Chen, R. Song, H. Tian et al., Graphene/semiconductor heterojunction solar cells with modulated antireflection and graphene work function, *Energy Environ. Sci.* 6, 2013, 108–115. doi: 10.1039/c2ee23538b
14. F. Jiao, H. Yen, G.S. Hutchings, B. Yonemoto, Q. Lu, F. Kleitz, Synthesis, structural characterization, and electrochemical performance of nanocast mesoporous Cu-/Fe-based oxides, *J. Mater. Chem. A.* 2, 2014, 3065–3071. doi: 10.1039/c3ta14111j

15. X. Miao, S. Tongay, M.K. Petterson, K. Berke, A.G. Rinzler, B.R. Appleton, A.F. Hebard, High efficiency graphene solar cells by chemical doping, *Nano Lett.* 12, 2012, 2745–2750. doi: 10.1021/nl204414u
16. Y. Shi, K.K. Kim, A. Reina, M. Hofmann, L.J. Li, J. Kong, Work function engineering of graphene electrode via chemical doping, *ACS Nano* 4, 2010, 2689–2694. doi: 10.1021/nn1005478
17. G. Li, R. Zhu, Y. Yang, Polymer solar cells, *Nat. Photonics* 6, 2012, 153–161. doi: 10.1038/nphoton.2012.11
18. S. Günes, H. Neugebauer, N.S. Sariciftci, Conjugated polymer-based organic solar cells, *Chem. Rev.* 107, 2007, 1324–1338. doi: 10.1021/cr050149z
19. W. Cai, X. Gong, Y. Cao, Polymer solar cells: Recent development and possible routes for improvement in the performance, *Sol. Energy Mater. Sol. Cells* 94, 2010, 114–127. doi: 10.1016/j.solmat.2009.10.005
20. T. Mahmoudi, Y. Wang, Y. Hahn, Graphene and its derivatives for solar cells application, *Nano Energy* 47, 2018, 51–65. doi: 10.1016/j.nanoen.2018.02.047
21. J. You, L. Dou, Z. Hong, G. Li, Y. Yang, Recent trends in polymer tandem solar cells research, *Prog. Polym. Sci.* 38, 2013, 1909–1928. doi: 10.1016/j.progpolymsci.2013.04.005
22. K. Cui, S. Maruyama, Multifunctional graphene and carbon nanotube films for planar heterojunction solar cells, *Prog. Energy Combust. Sci.* 70, 2019, 1–21. doi: 10.1016/j.pecs.2018.09.001
23. B.E. Hardin, H.J. Snaith, M.D. McGehee, The renaissance of dye-sensitized solar cells, *Nat. Photonics* 6, 2012, 162–169. doi: 10.1038/nphoton.2012.22
24. S.C. Ameta, R. Ameta, Dye-sensitized solar cells, in: *Solar Energy Conversion and Storage Photochemical Modes*, CRC press, Boca Raton 2015: pp. 85–113. doi: 10.1201/b19148
25. J.D. Roy-Mayhew, I.A. Aksay, Graphene materials and their use in dye-sensitized solar cells, *Chem. Rev.* 114, 2014, 6323–6348. doi: 10.1021/cr400412a
26. A. Kongkanand, K. Tvrdy, K. Takechi, M. Kuno, P. V Kamat, Quantum dot solar cells. Tuning photoresponse through size and shape control of CdSe−TiO_2 architecture, *J. Am. Chem. Soc.* 130, 2008, 4007–4015. doi: 10.1021/ja0782706
27. S. Rühle, M. Shalom, A. Zaban, Quantum-dot-sensitized solar cells, *ChemPhysChem.* 11, 2010, 2290–2304. doi: 10.1002/cphc.201000069
28. K.A.S. Fernando, S. Sahu, Y. Liu, W.K. Lewis, E.A. Guliants, A. Jafariyan, P. Wang, C.E. Bunker, Y.P. Sun, Carbon quantum dots and applications in photocatalytic energy conversion, *ACS Appl. Mater. Interfaces* 7, 2015, 8363–8376. doi: 10.1021/acsami.5b00448
29. H.S. Jung, N. Park, Perovskite solar cells: From materials to devices, *Small* 11, 2015, 10–25. doi: 10.1002/smll.201402767
30. T. Miyasaka, A. Kojima, K. Teshima, Y. Shirai, Organometal halide perovskites as visible-light sensitizers for photovoltaic cells, *J. Am. Chem. Soc.* 131, 2009, 6050–6051. doi: 10.1021/ja809598r
31. M.M. Lee, J. Teuscher, T. Miyasaka, T.N. Murakami, H.J. Snaith, Efficient hybrid solar cells based on meso-superstructured organometal halide perovskites, *Science* 338, 2012, 643–647. doi: 10.1126/science.1228604
32. S. Collavini, J.L. Delgado, Carbon nanoforms in perovskite-based solar cells, *Adv. Energy Mater.* 7, 2017, 1601000. doi: 10.1002/aenm.201601000
33. H. Li, K. Cao, J. Cui, S. Liu, X. Qiao, Y. Shen, M. Wang, 14.7% efficient mesoscopic perovskite solar cells using single walled carbon nanotubes/carbon composite counter electrodes, *Nanoscale* 8, 2016, 6379–6385. doi: 10.1039/C5NR07347B
34. H. Wei, J. Xiao, Y. Yang, S. Lv, J. Shi, X. Xu, J. Dong, Y. Luo, D. Li, Q. Meng, Free-standing flexible carbon electrode for highly efficient hole-conductor-free perovskite solar cells, *Carbon* 93, 2015, 861–868.

7 Fuel Cell

7.1 INTRODUCTION TO FUEL CELL TECHNOLOGY

The basic principle of the fuel cell was discovered by Christian Friedrich Schönbein in the year 1838. William Grove developed the first fuel cell in 1939 by accident, based on reversing the electrolysis of water [1]. Fuel cells are static energy conversion devices that directly convert the chemical energy of a reaction into electrical energy with high efficiency and low emission of pollutants and produce water and heat as by-products. A fuel cell includes close packing of the current collector, gas inlet/outlet, gaskets, gas diffusion layer and membrane electrode assembly (MEA). The different components of a fuel cell are presented schematically in Figure 7.1. MEA is the heart of the fuel cell, consists of two electrodes, cathode and anode, a proton exchange membrane and an electrolyte [2]. In the cathodic counterpart of the fuel cell, a reduction of oxygen from air takes place, whereas at the anodic part, oxidation of hydrogen occurs. The electrolyte membrane allows cation from the anode to cathode compartment, which is generated from the oxidation of H_2, but acts as insulation for the electron. The electron moves through the external circuit to the cathode. The recombination of positive and negative ion takes place at the anode and produces water.

$$\text{Anode reaction:} \quad H_2 = 2H^+ + 2e^- \tag{7.1}$$

$$\text{Cathode reaction:} \quad \frac{1}{2} 2H^+ + 2e^- = H_2O \tag{7.2}$$

$$\text{Overall reaction:} \quad H_2 + \frac{1}{2} O_2 = H_2O \tag{7.3}$$

The flow of ionic charge all the way through the electrolyte is required to balance by the flow of electronic charge through an exterior circuit, and this is the balance of electron that produces electricity [3].

Fossil fuel resources are being consumed at exceptional rates, and as a result, greenhouse gas emission of more than 31 billion tons of CO_2 occurs due to combustion procedures. The dream of the electric vehicle has a long history and has become real due to the unsustainable consumption of fossil fuels by transportation and its unbearable environmental consequences [4]. Fuel cell electric vehicles run on hydrogen gas instead of gasoline and thus do not depend on fossil oil. But fuel cell technologies have to compete with usual gasoline vehicles due to their high cost and insufficient performance durability [5]. It is clear that society requires such clean technologies, which are sustainably clean in service, are practically feasible, are high performing and efficient and most importantly are affordable [4,5].

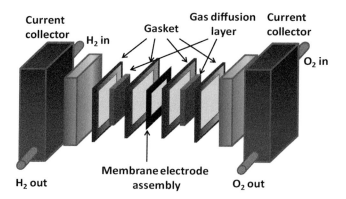

FIGURE 7.1 Different components of proton exchange membrane fuel cell.

The performance of the fuel cell depends on the anode and cathode catalyst. Efficient, economic and stable catalyst development plays a key role in the development of a fuel cell. A cathode electrode in the fuel cell typically consists of the catalyst and the conductive supporting material. The supporting layer generally acts as the diffusion layer as well as the current collector. A novel metal-based catalyst, such as Pt/C, has been extensively used as a cathode catalyst due to its high efficiency, durability and stability. Recently, highly active and stable precious-metal-free catalysts such as nanocarbon materials with different morphologies and properties have been vigorously used as cathode catalysts in the fuel cell. The cost of the fuel cell mainly depends on the cathode catalyst (47%–75% of the total cost). Therefore, the recent trend is focussed on the development of the carbon-based cathode catalyst to improve the cycling durability and resistance to catalytic poisoning with methanol, CO and sulfide.

7.2 WORKING PRINCIPLE AND MECHANISM

The working mechanism of the fuel strongly depends on the catalysts and electrolyte used in the fuel cell. The fuel cell consists of two electrodes and electrolyte connected with electrodes on either side. In anode electrode, the hydrogen is always fed, and in cathode the oxygen from air is fed continuously. Therefore, hydrogen fuel is decomposed into positive ions and negative ions at the anode end. The electrolyte membrane placed between two electrodes permits only the positive ions to flow from anode to cathode end and behave as an insulator for electrons. The electrons generated from the anode used then recombine on the other end of the membrane. and the free electrons move to the cathode end via an external electrical circuit. The recombination of the negative and positive ions with oxygen takes place at the cathode to form pure water as a by-product. The diagram for the mechanism of a fuel cell is shown in Figure 7.2.

The oxygen reduction reaction (ORR) of the fuel cell generally depends on the catalyst used and the pH of the electrolyte. The electrochemical reduction of oxygen in an aqueous (acidic) electrolyte can progress through two overall pathways – in acidic medium and in neutral or basic medium.

Fuel Cell

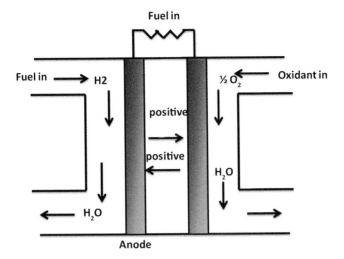

FIGURE 7.2 Diagram for the operation of fuel cell.

ORR pathway in acidic medium

Four-electron pathway:

$$O_2 + 4H^+ + 4e^- \rightarrow 2H_2O \qquad E = 1.229\,V \qquad (7.4)$$

Two-electron pathway:

$$O_2 + 2H^+ + 2e^- \rightarrow H_2O_2 \qquad E = 0.695\,V \qquad (7.5)$$

$$H_2O_2 + 2H^+ + 2e^- \rightarrow 2H_2O \qquad E = 1.770\,V \qquad (7.6)$$

$$2H_2O_2 \rightarrow 2H_2O + O_2 \qquad (7.7)$$

Recent studies advise that the ORR in a neutral or alkaline electrolyte proceeds dominantly via the OH^- producing the following pathway.

ORR pathway in neutral or basic medium

Four-electron pathway:

$$O_2 + 2H_2O + 4e^- \rightarrow 2OH^- \qquad E = 0.401\,V \qquad (7.8)$$

Two-electron pathway:

$$O_2 + H_2O + 2e^- \rightarrow HO_2^- + OH^- \qquad E = -0.065\,V \qquad (7.9)$$

$$HO_2^- + H_2O + 2e^- \rightarrow 3OH^- \qquad E = 0.867\,V \qquad (7.10)$$

$$2HO_2^- + H_2O + 2e^- \rightarrow 2OH^- + O_2 \qquad (7.11)$$

To check the ORR activity, a rotating disk electrode (RDE) has been used. The evaluation of a catalyst is based on the kinetic rates and reaction mechanism under non-mass-transfer limited conditions. The intrinsic property of the catalyst, which is kinetic current without mass transfer effect of the catalysts, is possible to derive by the Koutecky-Levich equation (Equation 7.12):

$$\frac{1}{j} = \frac{1}{j_k} + \frac{1}{j_d} \qquad (7.12)$$

$$j_d = 0.62\, nF\, C_{O_2} D_{O_2}^{2/3} \nu^{1/6} \omega^{1/2} \qquad (7.13)$$

where j is the limiting current density, j_k is the kinetic current density, j_d is the diffusion-limited current density, F is the Faraday constant (F = 96,485 C/mol) [6], C_{O_2} (1.2 × 10^{-6} mol/cm³) is the saturated concentration of oxygen in 0.1 M KOH [7,8], D_{O_2} (1.9 × 10^{-5} cm²/s) is the diffusion coefficient of oxygen in 0.1 M KOH [8,9], ν is the kinematic viscosity of 0.1 M KOH (n = 0.01 cm²/s) [9], and ω is the electrode rotation rate in rad/s. The number of electrons involved is calculated from the slope of the Koutecky-Levich equation. The number of electrons involved for commercial Pt/C, is approximately 3.96 [10], which suggests an oxygen reduction by Pt/C is a four-electron transfer process.

7.3 CLASSIFICATION OF FUEL CELL

According to the choice of electrolyte and fuel, fuel cells are classified into six major categories as follows (Table 7.1):

1. Proton exchange membrane fuel cell (PEMFC)
    a. Direct formic acid fuel cell (DFAFC)
    b. Direct ethanol fuel cell (DEFC)
2. Alkaline fuel cell (AFC)
    a. Proton ceramic fuel cell (PCFC)
    b. Direct borohydride fuel cell (DBFC)
3. Phosphoric acid fuel cell (PAFC)
4. Molten carbonate fuel cell (MCFC)
5. Solid oxide fuel cell (SOFC)
6. Direct methanol fuel cell (DMFC)

7.3.1 Proton Exchange Membrane Fuel Cell

The PEMFC is composed of a solid polymer electrolyte to exchange the ions between cathode and anode at the operating temperature which is as low as around 100°C. The membrane is a wonderful conductor of protons and an insulator in case of electrons. The chemical reactions in anode and cathode sides as well as their overall reactions are given in Equations 7.1–7.3. The high-power density and quick start-up

TABLE 7.1
Different Types of Fuel Cells and Their Features

Type	Temperature (°C)	Fuel	Electrolyte	Mobile Ion
Proton exchange membrane fuel cell (PEMFC)	100	H_2, CH_3OH	Sulfonated polymer (Nafion)	H^+
Alkaline fuel cell (AFC)	100–250	H_2	KOH	OH^-
Phosphoric acid fuel cell (PAFC)	175–200	H_2	H_3PO_4	H^+
Molten carbonate fuel cell (MCFC)	600–700	Hydrocarbon, CO	Na_2CO_3, K_2CO_3	CO_3^{2-}
Solid oxide fuel cell (SOFC)	1000	Hydrocarbon, CO	$(Zr,Y)O_{2-\delta}$	O^{2-}
Direct methanol fuel cell (DMFC)		H_2, CH_3OH	Polymer	H^+

for automotive vehicles are the advantages of the PEMFC. Two different types of PEMFCs are DFAFC and DEFC. DFAFC is the fuel cell, where inlet fuel formic acid (HCOOH) is directly fed to the anode electrode, but the ethanol as input fuel instead of hydrogen is used in DEFC.

7.3.2 Alkaline Fuel Cell

The AFC is one of the earlier fuel cell systems where an aqueous solution of the potassium hydroxide (KOH) as an electrolyte is used. The major disadvantage of the AFC is that it is sensitive to CO_2 because of more reaction time and consumes alkaline electrolyte, thereby reducing the concentration of hydroxide ion during chemical reactions [11].

7.3.3 Solid Oxide Fuel Cell

The SOFCs are fundamentally high-temperature fuel cells. In the SOFC dense yttria-stabilised zirconia is used as electrolyte, which is basically a solid ceramic material. The oxygen (O^{2-}) ions combine with hydrogen (H^+) ions and generate water and heat. The SOFC is used to generate electricity at a high operating temperature of about 1000°C.

7.3.4 Direct Methanol Fuel Cell

The DMFC is relatively new in comparison with the other fuel cells. The DMFC uses polymer electrolyte like the PEM fuel cell. However, liquid methanol or alcohol as fuel instead of reformed hydrogen fuel is used in DMFC. The anode captures hydrogen by dissolving liquid methanol (CH_3OH) in water in order to get rid of the external reformer. The recombination of the positive ions and negative ions takes place at the cathode, which is supplied through an external circuit from the anode, and it is combined with oxidized air to produce water as a by-product.

7.3.5 Phosphoric Acid Fuel Cell

The PAFC utilises liquid phosphoric acid as a fuel. It is very tolerant to impurities in the reformed hydrocarbon fuels, unlike other fuel cells. The chemical reaction involved in the PAFC is the same as in the PEM fuel cell where pure hydrogen is used as its input fuel. The PAFC works at about 175°C–200°C.

7.3.6 Molten Carbonate Fuel Cell

The MCFC consists of two conductive porous electrodes in contact with a molten carbonate cell. Owing to its internal reforming capability, MCFC separates the hydrogen from carbon monoxide fuel, and decomposition of hydrogen takes place through the water shift reaction to produce hydrogen. It operates at high temperatures of about 600°C–700°C.

7.4 LIMITATION OF METAL-BASED CATALYST

It is well known that platinum has been extensively used as a precious metal-based cathode catalyst for a long time as the overpotential for ORR reaction is substantially low. Although the Pt-catalyst has brilliant catalytic activity, its comparative high cost restricts its application. The cost per unit power can be reduced if platinum is replaced with another metal such as Fe- or Co-based catalyst.

Moreover, metal-based catalysts are generally vulnerable to unpleasant environmental conditions. The catalytic activity of Pt-catalyst is reduced in sulfide- or chloride-rich environments.

Metal-based catalysts are also sensitive to high cathodic pH values, which result in the crossover of cations through the membrane to the cathode compartment.

7.5 CARBON-BASED CATALYST

Nanocarbon materials combine all the requirements to be an ideal fuel cell catalyst. Carbonaceous materials are highly conductive, durable, have good mechanical strength and are inexpensive. Therefore, they are becoming the most promising alternatives to precious metal-based catalysts. In addition to the high catalytic activity of the nanocarbon-based catalyst, they show exceptional cycling stability and resistance to catalytic poisoning with CO, methanol, and sulfide. It is worth mentioning that using carbon-based catalysts has an advantage over metal-based catalysts in terms of cost effectiveness. Recently, researchers found a way to prepare carbon-based catalysts from sustainable precursors, including natural or waste materials. Activated carbon, graphene, hetero-atom-doped graphene and CNTs are being extensively used as non-metal catalysts [12]. A representation of a fuel cell using carbon-based material for both anode and cathode can be found in Figure 7.3.

7.5.1 Carbon-Based Anode Catalyst

The high-efficiency anode reaction (methanol oxidation) depends mainly on the performance of anode catalysts in terms of high activity, reliable stability and

Fuel Cell

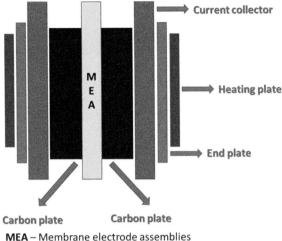

FIGURE 7.3 Diagram for a fuel cell with carbon-based catalysts as both anode and cathode.

good poison tolerance. It is true that the use of carbon-based catalysts is the main alternative route in this area. Different carbon-based materials such as graphene, carbon nanotubes, activated carbon, heteroatom-doped carbon, metal oxide–modified carbon and polymer-modified carbon supported catalysts have been used traditionally as anode catalyst.

Graphene and its composites have been proven for high electrocatalytic activity towards the methanol oxidation reaction [13–15]. It is reported that the mass activities of these Pt/graphene catalysts were at least twice as large as those observed with conventional Pt/C catalysts. Li and coworkers described the fabrication of Pt NPs decorated graphene sheets, which have larger electrochemically active surface area (ECSA) (36.3 m^2 g^{-1}) and better tolerance towards CO. The as-obtained Pt/graphene hybrid showed much more enhanced catalytic activity when compared with Pt/MWNT and other graphene-supported various nanostructures [5,3]. Kim et al. used thermal exfoliation of graphene nanosheets as highly functionalised materials for deposition of Pt NPs [16] and PtRu NPs [17] on graphene sheets with high loading, which could help in maintaining the highly active surface areas after several hours of catalyst use.

Carbon black and mesoporous carbon can be competent for a universal platform for loading metal NPs with controllable compositions and shapes, which is good for electrochemical methanol oxidation [18]. Carbon black decorated with PtRu electrocatalyst has been investigated for methanol oxidation reaction [19]. Su et al. described that the Pt/ordered MC catalyst holds a significantly higher ECSA (27.0 m g^{-1}), which is an excellent criterion for methanol electro-oxidation [20].

Carbon nanotubes (denoted as SWCNTs or MWCNTs, respectively) have shown very exciting results as catalyst support for fuel cell applications in the last decade [21,22]. Gattia and co-workers demonstrated loading of platinum clusters on three carbonaceous supports containing MWCNTs, single-wall carbon nanohorns (SWCNHs), and Vulcan XC-72 [23]. Electrochemical characterisation shows that platinum nanoparticles decorated on MWCNTs were much more proficient than those on other carbon material towards methanol oxidation reaction.

Each carbon nanomaterial has its own advantages and disadvantages as cathode catalyst support. Heteroatom-doped carbon nanostructures such as nitrogen, boron, sulfur, etc. can significantly affect the electronic structures, chemical reactivity, pH as well as conductivity of carbon nanomaterials and thus are good for DMFC catalyst. Recently, Sun and co-workers demonstrated a low-temperature (160°C) solvothermal route for the synthesis of nano-flower-like N-doped graphene [24]. Manthiram et al. synthesised B-doped CNTs by a chemical vapor deposition method using toluene and triethyl borate as the precursors for C and B [25]. However, Amini's group developed a synthesis route of S-doped CNTs, where impregnation of CNTs with sulfur from toluene solution followed by a melt-coating step was used [26].

7.5.2 Carbon-Based Cathode Catalyst

Metal-free nanocarbon-based catalyst is superior to Pt-based catalyst, because they encompass unique electronic structures, good conductivity and most importantly high surface area. Nanocarbon catalysts such as graphene, graphitic C_3N_4 (g-C_3N_4), CNTs, carbon black, mesoporous carbon and heteroatom-doped carbon have been traditionally used for oxygen reduction reaction. They not only show better ORR activity than the commercial Pt/C catalyst but also disclose an enhanced selectivity for oxygen reduction in the presence of methanol, thus avoiding a crossover effect [27].

Several studies reported the usage of graphene-supported precious platinum as efficient ORR electrocatalyst activity in fuel cell. It is well known that chemically reduced graphene oxide (rGO) nanosheets have better catalytic activity than GO, due to higher surface area, to better conductivity and maybe to some heteroatom like N and S attached to the rGO sheets, which may come during chemical reduction of GO [27]. N-doped graphene is the most studied among other nanocarbon catalyst as electrocatalyst. The reason behind good catalytic activity is that the lone electron pairs of N atoms can form a delocalised conjugated structure with the sp^2-hybridized carbon frameworks, thus resulting in an immense improvement in the electrocatalytic performance of graphene [28].

A variety of transition metal nanoparticles such as Mn_3O_4, Co_3O_4, $MnCo_2O_4$ and $Co_{1-x}S$ on the surface of rGO sheets and N-graphene have been incorporated or synthesised by a two-step method including coating of the metal precursors followed by hydrothermal reaction [10,29]. The synergetic effect of metal oxide/sulfide with graphene/N-graphene can lead to significantly advanced electrocatalytic activity of the hybrid material.

As a nanocarbon electrocatalyst, g-C_3N_4 has been occasionally explored as a cathode catalyst for applications in the fuel cell. As its semiconducting nature limits the electron transfer process, g-C_3N_4 composites with other carbonaceous material could show good catalytic activity in the fuel cell. Recently, Sun et al. [30] synthesised g-C_3N_4, a graphene in liquid-phase solution, using chemically converted graphene as a substrate that exhibited an enhanced electrocatalytic activity for ORR and CO tolerance as compared to that of platinum nanoparticles supported on graphene sheets.

Biomass holds a large amount of carbon in the forms of lignin, hemicelluloses and cellulose, and an easy calcinations process can translate it into carbon nanomaterials [31]. Recently, the utilisation of biomass and/or waste materials for the fabrication of ORR catalysts has attracted the research community due to its low cost and sustainable approach. Several biomass and waste materials such as seaweed, microorganisms, plant-mass, bacterial cellulose, waste wood and papers have been used to generate N-doped carbon for ORR electrocatalytic applications [32,33]. Liu et al. used rice straw for the preparation of N-doped carbon for ORR electrocatalyst [34]. Yuan et al. recently used the banana plant for the synthesis of N-doped carbon using the chemical activation of biochar, which also provides higher ORR catalytic activity [35].

7.6 CONCLUSION

This chapter systematically summarised different fuel cells as well as the performance and capabilities of different nanocarbon-based materials as efficient catalysts for fuel cell applications. Different types of fuel cells, their working principles and temperatures as well as mechanisms are explored in the first part of this chapter. Carbon materials are chemically more inert than metal-based ORR catalysts, although they provide long-term stability without significant loss of fuel cell performance because the rough surfaces of carbon materials and heteroatom doping help to enable its performance. Nanocarbon sometimes displays various surface functionalities, and the types of surface groups on carbon materials can differ the properties of ORR drastically. Metal oxides and polymers decorated with other nanocarbon materials can be superior catalysts for fuel cells. Although carbon-based fuel cell catalysts possess several advantages over metal-based catalysts, the large-scale production of carbon-based catalysts with a cost-effective route and environmentally friendly manner is still a challenge to researchers and requires more study.

REFERENCES

1. A. Boudghene Stambouli, E. Traversa, Fuel cells, an alternative to standard sources of energy, *Renew. Sustain. Energy Rev.* 6, 2002, 295–304. doi: 10.1016/S1364-0321(01)00015-6
2. A. Kirubakaran, S. Jain, R.K. Nema, A review on fuel cell technologies and power electronic interface, *Renew. Sustain. Energy Rev.* 13, 2009, 2430–2440. doi: 10.1016/j.rser.2009.04.004
3. Y. Li, W. Gao, L. Ci, C. Wang, Catalytic performance of Pt nanoparticles on reduced graphene oxide for methanol electro-oxidation, *Carbon N. Y.* 48, 2009, 1124–1130. doi: 10.1016/j.carbon.2009.11.034
4. D. Higgins, P. Zamani, A. Yu, Z. Chen, The application of graphene and its composites in oxygen reduction electrocatalysis: A perspective and review of recent progress, *Energy Environ. Sci.* 9, 2016, 357–390. doi: 10.1039/c5ee02474a
5. G. Wu, A. Santandreu, W. Kellogg, S. Gupta, O. Ogoke, H. Zhang, H.L. Wang, L. Dai, Carbon nanocomposite catalysts for oxygen reduction and evolution reactions: From nitrogen doping to transition-metal addition, *Nano Energy.* 29, 2016, 83–110. doi: 10.1016/j.nanoen.2015.12.032
6. Z.J. Lu, S.J. Bao, Y.T. Gou, C.J. Cai, C.C. Ji, M.W. Xu, J. Song, R. Wang, Nitrogen-doped reduced-graphene oxide as an efficient metal-free electrocatalyst for oxygen reduction in fuel cells, *RSC Adv.* 3, 2013, 3990–3995. doi: 10.1039/c3ra22161j

7. J. Qiao, L. Xu, P. Shi, L. Zhang, R. Baker, J. Zhang, Effect of KOH concentration on the oxygen reduction kinetics catalyzed by heat-treated co-pyridine/C electrocatalysts, *Int. J. Electrochem. Sci.* 8, 2013, 1189–1208.
8. R.E. Davis, G.L. Horvath, C.W. Tobias, The solubility and diffusion coefficient of oxygen in potassium hydroxide solutions, *Electrochim. Acta.* 12, 1967, 287–297. doi: 10.1016/0013-4686(67)80007-0
9. J. Zhang, L. Qu, G. Shi, J. Liu, J. Chen, L. Dai, N,P-codoped carbon networks as efficient metal-free bifunctional catalysts for oxygen reduction and hydrogen evolution reactions, *Angew. Chemie – Int. Ed.* 55, 2016, 2230–2234. doi: 10.1002/anie.201510495
10. H. Wang, J.T. Robinson, G. Diankov, H. Dai, Nanocrystal growth on graphene with various degrees of oxidation, *J. Am. Chem. Soc.* 132, 2010, 3270–3271. doi: 10.1021/ja100329d
11. M. Farooque, H.C. Maru, Fuel cells – The clean and efficient power generators, *Proc. IEEE.* 89, 2001, 1819–1829. doi: 10.1109/5.975917
12. M.J. Lázaro, S. Ascaso, S. Pérez-Rodríguez, J.C. Calderón, M.E. Gálvez, M.J. Nieto, R. Moliner et al. Carbon-based catalysts: Synthesis and applications, *Comptes Rendus Chim.* 18, 2015, 1229–1241. doi: 10.1016/j.crci.2015.06.006
13. D. Majumdar, Graphene-MnO_2 composite as electrocatalyst for oxygen, *Acad. J Aureole,* 4, 2016, pp. 31–37, ISSN No. 0976-9625.
14. S. Bong, Y.R. Kim, I. Kim, S. Woo, S. Uhm, J. Lee, H. Kim, Graphene supported electrocatalysts for methanol oxidation, *Electrochem. Commun.* 12, 2010, 129–131. doi: 10.1016/j.elecom.2009.11.005
15. J.C. Ng, C.Y. Tan, B.H. Ong, A. Matsuda, Effect of synthesis methods on methanol oxidation reaction on reduced graphene oxide supported palladium electrocatalysts, *Procedia Eng.* 184, 2017, 587–594. doi: 10.1016/j.proeng.2017.04.143
16. S.M. Choi, M.H. Seo, H.J. Kim, W.B. Kim, Synthesis of surface-functionalized graphene nanosheets with high Pt-loadings and their applications to methanol electrooxidation, *Carbon N. Y.* 49, 2010, 904–909. doi: 10.1016/j.carbon.2010.10.055
17. C. Nethravathi, E.A. Anumol, M. Rajamathi, N. Ravishankar, Highly dispersed ultrafine Pt and PtRu nanoparticles on graphene: Formation mechanism and electrocatalytic activity, *Nanoscale.* 3, 2011, 569–571. doi: 10.1039/c0nr00664e
18. Q. Jiang, L. Jiang, J. Qi, S. Wang, G. Sun, Experimental and density functional theory studies on PtPb/C bimetallic electrocatalysts for methanol electrooxidation reaction in alkaline media, *Electrochim. Acta.* 56, 2011, 6431–6440. doi: 10.1016/j.electacta.2011.04.135
19. E. Antolini, Carbon supports for low-temperature fuel cell catalysts, *Appl. Catal. B Environ.* 88, 2008, 1–24. doi: 10.1016/j.apcatb.2008.09.030
20. F. Su, C.K. Poh, Z. Tian, G. Xu, G. Koh, Z. Wang, Z. Liu, J. Lin, Electrochemical behavior of pt nanoparticles supported on meso- and microporous carbons for fuel cells, *Energy and Fuels.* 24, 2010, 3727–3732. doi: 10.1021/ef901275q
21. G. Girishkumar, T.D. Hall, K. Vinodgopal, P. V Kamat, Single wall carbon nanotube supports for portable direct methanol fuel cells, *J. Phys. Chem. B.* 110, 2006, 107–114. doi: 10.1021/jp054764i
22. R. Chetty, W. Xia, S. Kundu, M. Bron, T. Reinecke, W. Schuhmann, M. Muhler, Effect of reduction temperature on the preparation and characterization of Pt-Ru nanoparticles on multiwalled carbon nanotubes, *Langmuir.* 25, 2009, 3853–3860. doi: 10.1021/la804039w
23. D. Mirabile Gattia, M.V. Antisari, L. Giorgi, R. Marazzi, E. Piscopiello, A. Montone, S. Bellitto, S. Licoccia, E. Traversa, Study of different nanostructured carbon supports for fuel cell catalysts, *J. Power Sources.* 194, 2009, 243–251. doi: 10.1016/j.jpowsour.2009.04.058
24. D. Geng, Y. Hu, Y. Li, R. Li, X. Sun, Electrochemistry communications one-pot solvothermal synthesis of doped graphene with the designed nitrogen type used as

a Pt support for fuel cells, *Electrochem. Commun.* 22, 2012, 65–68. doi: 10.1016/j.elecom.2012.05.033
25. S. Wang, T. Cochell, A. Manthiram, Boron-doped carbon nanotube-supported Pt nanoparticles with improved CO tolerance for methanol electro-oxidation, *Phys. Chem. Chem. Phys.* 14, 2012, 13910–13913. doi: 10.1039/c2cp42414b
26. R. Ahmadi, M.K. Amini, Synthesis and characterization of Pt nanoparticles on sulfur-modified carbon nanotubes for methanol oxidation, *Int. J. Hydrogen Energy.* 36, 2011, 7275–7283. doi: 10.1016/j.ijhydene.2011.03.013
27. Y. Zheng, Y. Jiao, M. Jaroniec, Y. Jin, S.Z. Qiao, Nanostructured metal-free electrochemical catalysts for highly efficient oxygen reduction, *Small.* 8, 2012, 3550–3566. doi: 10.1002/smll.201200861
28. Q. Li, R. Cao, J. Cho, G. Wu, Nanocarbon electrocatalysts for oxygen reduction in alkaline media for advanced energy conversion and storage, *Adv. Energy Mater.* 4, 2014. doi: 10.1002/aenm.201301415
29. H. Wang, Y. Liang, Y. Li, H. Dai, Co1-xS-graphene hybrid: A high-performance metal chalcogenide electrocatalyst for oxygen reduction, *Angew. Chemie – Int. Ed.* 50, 2011, 10969–10972. doi: 10.1002/anie.201104004
30. Y. Sun, C. Li, Y. Xu, H. Bai, Z. Yao, G. Shi, Chemically converted graphene as substrate for immobilizing and enhancing the activity of a polymeric catalyst, *Chem. Commun.* 46, 2010, 4740–4742. doi: 10.1039/c001635g
31. X. Zhang, Q. Yan, J. Li, I.-W. Chu, H. Toghiani, Z. Cai, J. Zhang, Carbon-based nanomaterials from biopolymer lignin via catalytic thermal treatment at 700 to 1000°C, *Polymers (Basel).* 10, 2018, 183. doi: 10.3390/polym10020183
32. H. Liang, Z. Wu, L. Chen, Bacterial cellulose derived nitrogen-doped carbon nano fiber aerogel: An efficient metal-free oxygen reduction electrocatalyst for zinc-air battery, *Nano Energy.* 11, 2015, 366–376. doi: 10.1016/j.nanoen.2014.11.008
33. H. Zhu, J. Yin, X. Wang, H. Wang, X. Yang, Microorganism-derived heteroatom-doped carbon materials for oxygen reduction and supercapacitors, *Adv. Funct. Mater.* 23, 2013, 1305–1312. doi: 10.1002/adfm.201201643
34. L. Liu, Q. Xiong, C. Li, Y. Feng, S. Chen, Conversion of straw to nitrogen doped carbon for efficient oxygen reduction catalysts in microbial fuel cells, *RSC Adv.* 5, 2015, 89771–89776. doi: 10.1039/c5ra15235f
35. H. Yuan, L. Deng, Y. Qi, N. Kobayashi, J. Tang, Nonactivated and activated biochar derived from bananas as alternative cathode catalyst in microbial fuel cells, *Sci. World J.* 2014, 2014. doi: 10.1155/2014/832850

8 Electrocatalysis and Photocatalysis

8.1 INTRODUCTION

The two main problems and challenges of the modern world are increasing energy demand and global warming due to the increase in the carbon dioxide (CO_2) concentration in the atmosphere. Because of exponential population growth, increasing demands and limited availability of fossil fuels, the world is facing an energy crisis. Also, hackneyed use of fossil fuels leads to global warming and escalates environmental pollution [1]. For the purposes of the fuel cell, currently, the burning of fossil fuels for steam reforming and water electrolysis [$2H_2O(l) \rightarrow 2H_2(g) + O_2(g)$: $\Delta G° = +237.2$ KJ mol^{-1}, $\Delta E° = 1.23$ V (T = 25°C, P = 1 atm) versus reversible hydrogen electrode (RHE)] is done for the production of H_2. The problem is that the burning of fossil fuels is not suitable for the environment because it generates CO_2 along with H_2 that compromises the purity of H_2 as well as leads the world towards global warming. The world is facing many problems, but energy and global warming is one of the problems that urgently needs to be addressed. Researchers are working in the direction to solve these two major issues by adopting various methods. Electrocatalysis [2,3], photocatalysis [4] and photoelectrocatalysis [4] are the efficient ways by which we can generate H_2 fuel by splitting water into hydrogen and oxygen and also reduce CO_2 in a sustainable direction. In the last few decades, tremendous research has been done to investigate carbon-based material as a catalyst in energy generation and conversion because of its ease of production, abundance and excellent electrical, mechanical and chemical properties [5,6]. Various practices for the involvement and incorporation of carbon-based nanomaterials in the fields of water splitting and CO_2 reduction have been performed in recent years to utilise its extraordinary properties.

Different derivatives of carbon such as graphene [5,6], carbon nanotubes (CNTs) [7–9], graphitised carbon nitrides (g-C_3N_4) [10], carbon nanowires (CNWs), etc. along with its doping with heteroatoms have been explored for catalysis and proved to deliver interesting and satisfactory results. But, there are still challenges to overcome in this particular research field. However, carbon remains attractive to researchers because of its environmentally friendly nature.

In the electrocatalysis of water splitting, water splits into hydrogen (H_2) and oxygen (O_2) by means of an electric current in the presence of a catalyst and an electrolyte. Some kinds of membranes are used to separate both the gases, avoid any unwanted reaction from taking place and permit only ions to flow through them [10]. Carbon is present abundantly in nature and has a very important role in human life. Several research efforts have been made over the past few decades in the synthesis and application of carbon-based materials. Carbon possesses a multidimensional nanoarchitecture like 0D (fullerenes), 1D (carbon nanotubes), 2D (graphene) and

3D (graphite) with tunable chemical, electronic and physical characteristics. And these prominent qualities of carbon are well utilised in the field of catalysis for water splitting. The exploration of carbon has gone through several phases and basically started from the discovery of a new allotrope of carbon, buckminsterfullerene C_{60} in 1985, and in 1991, carbon nanotubes were invented [10]. Breakthrough came in the year 2004, when a one-atom-thick, sp^2-hybridized carbon called *graphene* was discovered [10]. There are several carbon-based materials such as metal-free carbon, graphene, metal organic frameworks (MOFs), metal-doped carbon, conjugated polymers, etc. that are fruitfully taking part in the field of electrocatalysis. However, different configurations of hybridisation lead to the different forms of carbon and decide their unique properties. Graphite, diamond and amorphous carbon are the three main classes of carbon based on their hybridisation. But instead of tremendous research in this field, their catalytic activities are far down from the noble metal-based catalyst. Presently, extensive research is going on in the search for a better and more efficient catalyst for water splitting and CO_2 reduction, for example, metal oxides, metal carbides, metal nitrides, metal phosphides, metal phosphates, metal alloys, layered double hydroxides (LDHs), metal sulphides, etc. Utilising the porous structure, low cost, high surface area, outstanding electrical conductivity and integrated structural as well as chemical stability, carbon and carbon-derived nanomaterials have been reported to be excellent catalysts for water splitting and CO_2 reduction. Carbonaceous material like g-C_3N_4, graphene and carbon nanotubes (CNTs) are the major candidates utilised in electrocatalysis. However, pristine carbon-based electrocatalyst does not show significant performance in their catalytic activity therefore induction of defects, doping with heteroatoms, forming composites, etc. practises have been done in order to integrate active sites to attain better activity of the catalyst. DFT calculations and computational analysis are also playing an important role for the analysis of structure and properties of carbon-based materials in electrocatalysis.

Photocatalysis with carbon-based materials is showing a vital growth towards the development of cost-effective and eco-friendly energy generation and conversion systems. Due to several advantages related to the carbon-derived catalyst, it has become a matter of exploration to meet future energy demands. Despite the fact that a lot of research has been focussed on the efficient activity of the catalyst to achieve high solar-to-hydrogen conversion (STH) and efficient CO_2 reduction, performance is still limited due to low absorption coefficients, poor quantum yields, etc. Recently, many carbon-based photocatalysts have been explored. Carbon-based materials such as graphene, graphene oxide (GO), polyimide polymers, carbon nitride (C_3N_4), holey graphene (C_2N), carbon quantum dots and their derivatives have been introduced in photocatalysts. Graphene is a metal free material derived from carbon containing graphite, which is earth abundant. To exploit the excellent properties of graphene, it can be coupled with another photocatalyst to achieve activity that is comparable to that of the noble metal catalyst [11]. Graphene acts as an excellent charge transporter because of its excellent thermal/electrical conductivity as well as mobility along with its ability to act as an excellent absorber. However, graphene without any functionalization and doping acts as a good conductor, so in order to design and tailor it as a photocatalyst, several efforts like composite formation, functionalization, heteroatom doping, etc. towards tuning its electronic properties that are suitable for photocatalysis have been practised. Another class of carbon-derived catalyst is g-C_3N_4, which is thermally and chemically stable and has excellent band tunability and possesses ideal band positions, a bandgap of 2.7 eV, having the top of the valence band at 1.6 eV and the bottom of the conduction band at −1.1 eV. But as stated previously, pristine carbon-derived catalyst is not a good and efficient catalyst. TiO_2 is known to be the best photocatalyst, and the doping of g-C_3N_4 with TiO_2 improved the catalytic performance [12]. The exfoliation (mechanical, chemical, hydrothermal and sonication) of the catalyst results in improved surface area, and as a result the electrical conductivity also increases. Doping with elements or suitable precursors of g-C_3N_4 is a very common practice to enhance catalytic activity. Other classes of carbon-based materials that are photochemically active catalysts are CNTs, carbon nanodots, carbon nanowires (CNWs), carbon quantum dots fullerenes (C_{60}) and boron carbides. Utilising the different qualities of carbon-based catalyst and its doping agents together, researchers are providing a new dimension in the development of much more efficient catalysts to reach sustainable and efficient energy pathways.

Various carbon-based nanomaterials such as nanotubes, nanowires and nanorods, as well as nanosheets, after heteroatom and metal doping show a significant catalytic activity for photoelectrocatalysis (PEC). The electronic properties of carbon nanomaterials can be tailored by tuning the energy bandgap by incorporating defects into the inert carbon materials. The incorporation of nanomaterials in semiconductors reduces the bandgap and allows more solar light absorption. Nanostructures can efficiently transport charges and can inject electrons and holes at the water-nanomaterial interface for oxidation and reduction processes. However, the efficiency of the overall PEC cell for water splitting is still quite low, and it is necessary to improve the efficiency and practicality of the photo-water splitting system, particularly by developing high-efficiency and cost-effective materials. The unique properties of nanomaterials provide great opportunities for developing systems with high photocatalytic efficiency for water splitting using sunlight.

8.2 WATER SPLITTING

Developing a clean, green and renewable energy source alternative to fossil fuel is an urgent demand. Among the various energy sources, hydrogen ($\Delta G = +237$ kJ mol^{-1}) as a fuel is gaining much attention because of its clean energy nature, its development towards a sustainable energy system and its potential to eliminate traditional energy sources. The generation of H_2 from renewable energy sources is versatile (Figure 8.1). So, efficient generation and storage of hydrogen are major challenges that need to be addressed. Natural resources such as water and sunlight are abundant and available supplies, which could be employed to split water for the production of chemical energy such as hydrogen (H_2). The generated H_2 gas is utilised in fuel cells. Fuel cell technology research is at its apex and is the best way to produce energy in an eco-friendly direction, as only H_2 and O_2 act as a fuel in the presence of an electrocatalyst and produce water as a by-product after combustion. This hydrogen generation strategy can be done through splitting the water by either the use of direct sunlight or by electricity to meet the low-cost production of H_2. Sunlight is an infinite and incessant source of energy; approximately 4.3×10^{20} J of sun energy

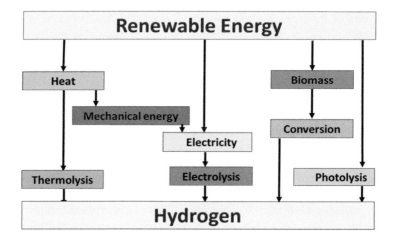

FIGURE 8.1 Flow diagram of hydrogen production from renewable energy sources.

strikes the earth's surface per hour, which is more than the current consumption of energy by mankind in 1 year [2]. In nature, water splitting is performed by plants in the presence of sunlight and is called *photosynthesis*, but hydrogen is not produced there. The water-splitting reaction consists of two half-reactions; that is, reduction of water which is an hydrogen evolution reaction (HER) and water oxidation which is an oxygen evolution reaction (OER), schematically shown in Figure 8.2.

The general reaction for water splitting is

$$2H_2O + 4h^+ \rightarrow O_2 + 4H^+ \quad \text{(OER at anode)}$$

$$4H^+ + 4e^- \rightarrow 2H_2 \quad \text{(HER at cathode)}$$

$$2H_2O + \text{Energy} \rightarrow 2H_2 + O_2 \quad \text{(Overall)}$$

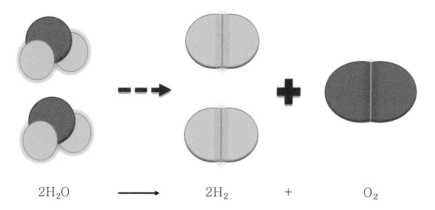

FIGURE 8.2 Water splitting into hydrogen and oxygen.

To fulfil this, the development of efficient, robust and economical catalysts is required for the splitting of water over a long period. H_2 produced from renewable energy sources and carbon neutral processes is known to be one of the most important sustainable energy generation routes to meet our future energy demands.

The splitting of water is realised by several methods:

- Electrolysis
- Photoelectrochemical water splitting
- Photocatalytic water splitting
- Radiolysis
- Thermal decomposition of water
  - Nuclear-thermal
  - Solar-thermal

Among these methods to split water, electrocatalysis, photocatalysis and photoelectrocatalysis methods are very economical and environmentally friendly, as they basically harvest renewable energy such as solar power.

8.2.1 Electrochemical Water Splitting

Electrolysis of water is a chemical method to split water into hydrogen (H_2) and oxygen (O_2) with the help of electricity. In 1789, this phenomenon was first observed by Nicholson and Carlisle and from then to the twentieth century more than 400 industrial water electrolysis units were in action; in 1939, a water electrolysis plant with a volume of 10,000 N m^3 h^{-1} H_2 became operable. From the journey of the Nicholson and Carlisle water electrolyser to the development of proton exchange membranes, the water-splitting phenomenon has gone through several architectures and mechanisms.

The chemical reactions taking place during water electrolysis are

$$2H^+(aq) + 2e^- \rightarrow H_2(g) \quad \text{(HER)}$$

$$2H_2O(l) \rightarrow O_2(g) + 4H^+(aq) + 4e^- \quad \text{(OER)}$$

For 100% faradic efficiency, ideally the production of hydrogen (2e$^-$ process) must be double that of oxygen (4e$^-$ process), and both H_2 and O_2 generation are proportional to the amount of charge species present in the electrolyte. A schematic illustration of a water electrolyser using platinum electrodes and graphite electrodes is shown in Figure 8.3.

It is interesting to know that decomposition of water through electrolysis was first performed in acidic water, and despite this, alkaline water is preferred in industry because of corrosion control. For the separation of hydrogen and oxygen gas during water electrolysis, different types of membrane are utilised, basically proton exchange membranes are used (Figure 8.4). Separating membranes is an important part of an electrolyser that allows only ions to pass through and not the molecules of H_2 and

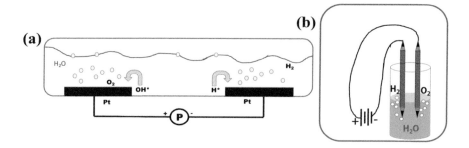

FIGURE 8.3 Water splitting (a) using platinum as an electrode and (b) using a graphite rod as an electrode.

O_2 and that keeps both gases separate to avoid the risk of explosion due to unwanted reactions. The thermodynamics and mechanisms of HER and OER in different media are discussed later.

The HER process undergoes two different mechanisms with three possible reaction steps in acidic media [13]. The first step is known as the discharge process, which is also called the Volmer reaction:

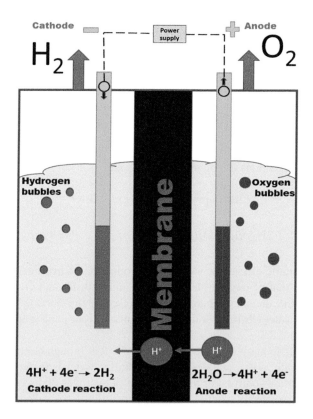

FIGURE 8.4 Full water electrolyser with membrane as a separator.

1. For capturing a proton, an electron is transferred to the surface of the catalyst which results in the formation of adsorbed hydrogen atom (H_{ads}) on the catalyst active sites.

$$H^+(aq) + e^- \rightarrow H_{ads} \quad \text{(Step 1, Volmer reaction)}$$

2. The Volmer step has two different pathways to generate H_2: one is known as Heyrovsky reaction pathway, and another is called Tafel reaction pathway.

The Heyrovsky reaction pathway is also known as the desorption step which takes place when the H_{ads} coverage on the catalyst surface is low. Due to low coverage, the adsorbed hydrogen atom prefers to combine with another proton and a new electron for the evolution of H_2:

$$H_{ads} + H^+(aq) + e^- \rightarrow H_2(g) \quad \text{(Step 2a, Heyrovsky)}$$

The Tafel reaction pathway for the generation of hydrogen is also known as a chemical desorption step which takes place at a relatively high coverage of H_{ads}. Because of the high coverage, recombination between the adjacent adsorbed hydrogen atoms is predominant:

$$H_{ads} + H_{ads} \rightarrow H_2(g) \quad \text{(Step 2b, Tafel)}$$

The HER reaction takes place differently in alkaline media due to its high pH values.

Similar to the steps of the HER process in acidic media, here also the reaction undergoes Volmer, Heyrovsky and Tafel reactions.

1. Since the concentrations of protons are less in alkaline media, instead of H^+, molecular H_2O combines with an electron to result in an adsorbed hydrogen atom on the surface of the catalyst. This reaction is called the Volmer reaction:

$$H_2O(l) + e^- + X \rightarrow XH_{ads} + OH^-(aq) + 4e^- \quad \text{(Volmer reaction)}$$

2. From the Volmer reaction for the generation of H_2, the reaction undergoes two steps: Heyrovsky and Tafel reactions.

In the Heyrovsky reaction, the adsorbed hydrogen atom couples with an electron and molecular H_2O to generate H_2 gas.

$$XH_{ads} + H_2O(l) + e^- \rightarrow H_2 + OH^- + X \quad \text{(Heyrovsky reaction)}$$

The Tafel reaction is the same as in the case of HER in acidic media, where two adsorbed hydrogen atoms combine to form H_2:

$$XH_{ads} + XH_{ads} \rightarrow H_2(g) + 2X \quad \text{(Tafel reaction)}$$

where X represents the site for hydrogen adsorption and XH_{ads} represents adsorbed hydrogen at the site.

HER in neutral media begins with water, and here also the reaction is first initiated by the Volmer reaction followed by the Heyrovsky and Tafel recombination steps [14,15].

$$H_2O(l) + e^- + X \rightarrow XH_{ads} + OH^-(aq) \quad \text{(Volmer reaction)}$$

$$H_2O(l) + XH_{ads} + e^- \rightarrow H_2 + OH^- + X \quad \text{(Heyrovsky reaction)}$$

$$2H_{ads} \rightarrow H_2(g) + 2X \quad \text{(Tafel reaction)}$$

where X stands for the catalytic site for hydrogen adsorption on the catalyst, and H_{ads} is hydrogen adsorbed on the surface of the catalyst.

The thermodynamics of OER is similar in both acidic and alkaline media [16]. For practical purposes, the overpotential exceeds the thermodynamically obtained potential (1.23 V). Apart from the novel metal catalyst for OER, all the other catalysts generally involve the adsorption of both O and OH groups on the surface of the catalyst.

The equations involved are

$$OH^- + * \rightarrow OH_{ads} + e^-$$

$$OH_{ads} + OH^- \rightarrow O_{ads} + H_2O + e^-$$

where * is the active site of the catalyst, and O_{ads} is the adsorbed oxygen on the catalyst surface. There are basically two pathways for the production of O_2 from the oxygen that is adsorbed on the catalytic active sites.

The first pathway is accomplished when the two O_{ads} couple with each other:

$$O_{ads} + O_{ads} \rightarrow O_2$$

And in the second step the O_{ads} reacts with OH^- to form an intermediate OOH_{ads} species that finally combines with OH^- to generate O_2:

$$O_{ads} + OH^- \rightarrow OOH_{ads} + e^-$$

$$OOH_{ads} + OH^- \rightarrow O_2 + H_2O + e^-$$

A similar trend follows for the generation of oxygen in neutral media, but the Volmer reaction begins with water here:

$$2H_2O + * \rightarrow OH_{ads} + H_2O + e^- + H^+$$

$$OH_{ads} + H_2O \rightarrow O_{ads} + H_2O + e^- + H^+$$

$$O_{ads} + H_2O \rightarrow OOH_{ads} + e^- + H^+$$

$$OOH_{ads} \rightarrow O_2 + e^- + H^+$$

where * is the active site of the catalyst, and O_{ads} is the adsorbed oxygen on the catalyst surface.

There are a few parameters that are used to evaluate the catalytic activity of the electrocatalyst based on the activity and efficiency, they include overpotential, Tafel slope, exchange current density, stability, faradic efficiency and turnover frequency. We discuss these terms in brief in upcoming sections.

Electrolysis of water can never take place at a theoretically determined potential. The difference between the experimental potential and the thermodynamically determined reduction (H_2, $E_O = 0.00$ V versus RHE)/oxidation (O_2, $E_O = 1.23$ V versus RHE) potential is known as overpotential (η) [17–19]. For more clarification, the fundamental requirement of a water-splitting system (electrolyser) is equilibrium potential $E_{eq} = 1.23$ V. Thus, a voltage of 1.23 V must be supplied to the electrolyser to generate H_2 at a specific rate, but operational voltage also depends upon the rate of kinetics of the water-splitting reaction between catalyst and electrolyte as well as design of the electrolyser unit:

$$E_{OP} = E_{eq} + \eta_A + \eta_C + \eta_\Omega$$

where η_A is the overpotential required to overcome the kinetic barrier at anode (OER), η_c is the overpotential required to overcome the kinetic barrier at cathode (HER) and η_Ω is the overpotential due to some additional resistances such as contact and solution resistances. The η_Ω drop in the voltage is known as the iR drop, which must be subtracted from the experimentally calculated potential.

Platinum (Pt) is a well-known catalyst for the generation of H_2, and ruthenium/iridium (Ru/Ir) is a catalyst for the generation of O_2 with almost zero overpotential. But there are some issues associated with the noble metals–based electrocatalyst – that is, their limited availability and high cost that contribute to the excessive production costs of fuel cell systems. Therefore, for the electrolysis of water, researchers are tailoring and designing a non-noble, non-precious electrocatalyst for efficient water-splitting reactions. For 10% efficient solar-to-fuel conversion devices, the expected current density (j) is 10 mA cm^{-2}. So, 10 mA cm^{-2} is the standard figure of merit at which usually the overpotential (η_{10}) is calculated [20,21]. For some materials, because of a dominant redox peak and current density more than 10 mA cm^{-2}, the overpotential at 50 mA cm^{-2} or 100 mA cm^{-2} is calculated [22]. This current density is calculated from a technique called *linear sweep voltammetry* (LSV, polarization curve) measured from galvanostat/potentiostat.

The relationship between the current (i) and voltage (V) is a fundamental concern in electrochemical reactions. In fact, varying the voltage changes the electrode kinetics with change in the interfacial current. For a basic understanding of the electrocatalytic reaction, Tafel plot analysis is a foremost concern. The Tafel slope and Tafel equation come from the Butler-Volmer equation [23]. During the electrochemical reaction, either the anodic or cathodic current dominates when the current increases:

$$i = i_0 \left\{ \exp\left[\frac{\alpha_{anodic} nF\eta}{RT}\right] - \exp\left[\frac{\alpha_{cathodic} nF\eta}{RT}\right] \right\} \quad \text{(Butler-Volmer equation)}$$

$$i = i_{anodic} + i_{cathodic}$$

where i_0 is the exchange current, which is the current at zero overpotential that depicts the rate of forward and reverse reaction when net electrolysis is zero; α_{anodic} is the charge transfer coefficient for the anodic reaction; $\alpha_{cathodic}$ is the charge transfer coefficient for the cathodic reaction; η is the overpotential; F is Faraday's constant; R is the gas constant; and T is temperature.

In near equilibrium both the anodic and cathodic terms are important, but farther from equilibrium one term dominates that can be analysed by the Tafel equation. We are considering the OER reaction, so we are taking anodic current as a candidate for the Tafel equation:

$$i = i_0 \left\{ \exp\left[\frac{\alpha_{anodic} nF\eta}{RT}\right] \right\}$$

Now, in logarithmic terms current will be

$$\log(i) = \log(i_0) + \eta/b \quad \text{(Tafel equation)}$$

$$\eta = a + b \log(i)$$

where a is constant (exchange current density), and b is the Tafel slope given by

$$b = \frac{\partial \eta}{\partial \log i} = 2.303 \, RT/\alpha_a F \quad \text{(Tafel slope)}$$

From the Tafel slope equation, the linear relationship between current log(i) and overpotential η can be predicted; hence, the plot between both can be drawn from the polarisation curve (LSV) which is known as the Tafel plot. The exchange current density (i_0) is determined by extrapolating the linear part of the Tafel plot and its intersection with the x-axis (Figure 8.5).

From this equation, we can see that the Tafel slope is inversely proportional to the transfer coefficient (α). Thus, the catalyst that has the high charge transfer coefficient has a small Tafel slope. And that is the reason why Tafel is a prime parameter in the electrocatalysis of water.

For better evaluation of the electrocatalyst, stability is a major parameter for splitting water at the industrial scale. The stability of the catalyst is generally accompanied by cyclic voltammetry (CV) scan at higher current density to test its degradation. In CV, the cycling is repeated hundreds or thousands of times in the active region of the

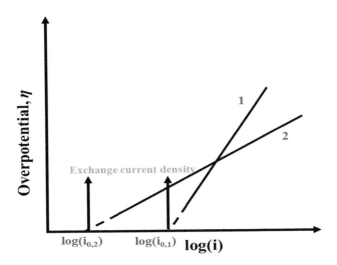

FIGURE 8.5 Graph depicting Tafel plot and exchange current density.

catalyst including the onset value. Another method for testing the stability is use of the chronoamperometric (at constant potential) or chronopotentiometric (at constant current) technique for more than 12 hours. After the degradation test, the shift in the overpotential at a certain current density (mainly 10 mA cm^{-2}) is compared to the initial overpotential. For a catalyst to be robust, efficient and stable, the shift in the overpotential at a particular current density should be minimal.

The efficiency by which the electrons provided by external circuit are capable enough to pilot the reaction for water splitting is called *faradic efficiency*. The ratio between the theoretical and experimental amounts of produced H_2/O_2 is needed to calculate faradic efficiency. The comparison is done because along with the desired gas, some faradic losses may occur due to the generation of heat and by-products during the reaction.

The turnover frequency (TOF) of the electrocatalyst is the measure of the number of moles of H_2/O_2 produced per unit time. This is a very important quantitative parameter for the analysis of electrocatalytic behaviour. The general formula for TOF is expressed as

$$TOF = IN_A/AFn\Gamma$$

where I is current, N_A is the Avogadro constant, A is the geometrical surface area, n is the number of electrons transferred, and Γ is the total concentration of catalyst (number of atoms).

Doping is the best practice to overcome the inertness of pristine carbon-based electrocatalyst in water splitting. The doping is accompanied by heteroatoms such as N, P, S or B to tailor the surface as well as electronic properties of the carbon catalyst that lead to the active sites' formation in the catalyst for the generation of hydrogen and oxygen gas [24]. Heteroatom-doped carbon nanomaterials have proven to be the best electrocatalysts for HER and ORR (oxygen reduction reaction), but they are not explored much in OER. Metal sulphides (Mo, Ni, Co, Fe/S) are the very prominent catalyst for water splitting. So, integrating these metal sulphides with carbon-based materials owing to their superior porous as well as electronic structures resulted in improved catalytic activities [13]. Mostly the catalytic activity of MoS_2 is high because of its rich active sites and better catalytic activities. So, the hybridisation of carbon-based nanomaterials (graphene, carbon nanotubes, nanofibers etc.) has been practised with MoS_2 to gain the improved catalytic activities [25–28]. Graphene-derived aerogels which are proven to have excellent porous structure and good electrical conductivity are better integrated with MoS_2. Liu et al. formed MoS_2 nanoparticles into mesoporous graphene foam (MGF) and that catalyst have showed a very efficient performance for HER that include a very low overpotential and very high stability because of the highly stable nature of MGF [29]. g-C_3N_4 is a polymeric material that is a semiconductor in nature having tri-s-triazine(melem) as the basic building unit connected by amino groups. Its bandgap is 2.7 eV with the band positions favourable for water splitting. It is worth mentioning that even in the absence of any metallic co-catalyst, g-C_3N_4 effectively produces H_2 gas by water

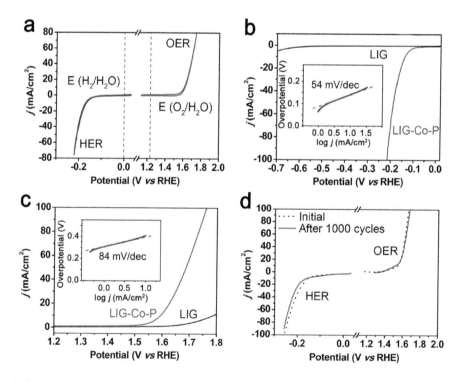

FIGURE 8.6 Electrocatalytic performance of the bifunctional LIG-Co-P. (a) Cyclic voltammetry curves for full water splitting of LIG-Co-P. The linear sweep voltammetry (LSV) of the LIG-Co-P for (b) H_2 evolution reaction (HER) and (c) oxygen evolution reaction (OER). Inset: corresponding Tafel plots. (d) Stability test (LSV) of the catalyst before and after 1,000 cycles for HER and OER. (Reproduced with permission from J. Zhang et al., *ACS Appl Mater Interfaces*, 9, 2017. doi: 10.1021/acsami.7b06727.)

splitting. Also, the thermodynamic driving force for OER using g-C_3N_4 is low. In order to design a metal free catalyst, g-C_3N_4 is combined with other carbonaceous materials (carbon dots/quantum dots) to achieve improved catalytic activities [30,31].

Zhang et al. reported a laser-induced graphene (LIG) doped with cobalt phosphide (LIG-Co-P) for electrochemically splitting water into hydrogen and oxygen [32]. The electrodes were fabricated on opposite faces of a plastic sheet. The high porosity and excellent conductivity of LIG help to split the water efficiently with long-term stability (Figure 8.6a–d).

Also, carbon is known to be a good choice as a substrate. Mostly electrocatalysts are grown on a substrate hydrothermally, using the chemical vapor deposition method, heat treated or electrochemically treated to avoid the use of binder (that reduces the stability), to inherit the conductivity of the substrate and to reduce the steps of the reaction. Mostly nickel foil, nickel foam, titanium foil, graphite plate and carbon plate are used as substrates. Carbon-based substrates are widely used because of their unique advantages along with cost effectiveness [33]. Suitable support of the catalyst prevents the agglomeration of the nanoparticles on the substrate, provides efficient charge transfer between the catalyst and substrate interfaces and provides uniformity in the nanoarchitecture of the deposited catalyst. Various metal phosphides, metal sulphides, metal selenides, sulfur – selenides hybrids etc. – were deposited on carbon cloth as a substrate [34]. Carbon nitrides, such as g-C_3N_4 and C_2N, also act as a promising substrate. Along with that, B_4C is a good candidate for the substrate because the dispersed nanoparticles on the substrate are uniform [33].

8.2.2 Photochemical Water Splitting

Photocatalytic water splitting is an artificial photosynthesis process that is done in a photochemical cell in the presence of an artificial or a natural light source. Here, the system is very simple, because only light, catalyst and water are required to split water into H_2 and O_2. The advantages associated with the photocatalysis of water splitting using light are shown schematically in Figure 8.7.

During photocatalysis, water splits into hydrogen and oxygen using photocatalyst (semiconductors) by capturing the solar energy invisible spectrum of light (380–700 nm) to generate H_2 fuel for a cost-effective, environmentally friendly route for energy generation.

The overall photocatalytic water-splitting reaction occurs mainly in three stages: (1) generation and separation of electron hole pair when light interacts, (2) migration of generated electron hole pair toward the surface of the catalyst and (3) generation of H_2 or O_2 because of the chemical reactions taking place at the surface of the catalyst (Figure 8.8). A suitable catalyst must be selected for the efficient generation of H_2 and O_2 during photocatalysis. All of the above steps are crucial in determining the overall performance and efficiency of the photocatalyst.

There are so many factors that determine the efficiency of H_2 produced during photocatalysis: light harvesting, separation of charges, transportation of charges, photon adsorption and oxidation and reduction at the surface are the steps that effect the photocatalytic efficiency. In this regard, tailoring the bandgap, hybridising the catalyst, coupling with narrow bandgap semiconductors, etc. are practises that

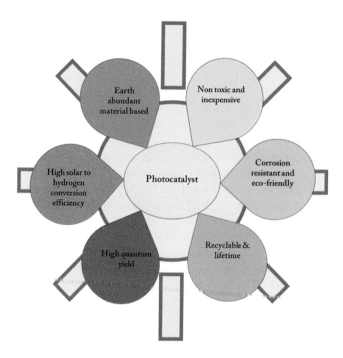

FIGURE 8.7 Advantages of catalysis utilising light.

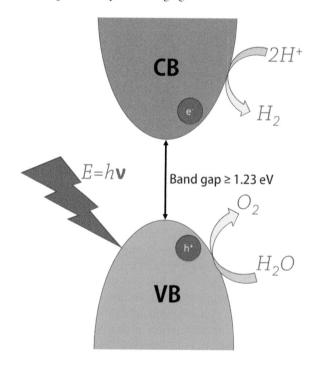

FIGURE 8.8 The role of the valance band and the conduction band in photocatalysis.

Electrocatalysis and Photocatalysis

have been done to achieve the best performance of the photocatalyst. What actually happens at the semiconductor-electrolyte interface?

- At a particular frequency of the incident light (greater than the bandgap energy), the electron hole pair separates and travels to the valance band (VB) and conduction band (CB).
- A photocurrent is generated due to unequal distribution of the migrated charge carrier (e^-/h^+) that disrupts the equilibrium.
- The oxidation (H_2O to O_2) and reduction (H_2O to H_2) of water take place at valance band and conduction band, respectively, as a result of photogenerated electrons (Figure 8.9).

$$2H^+ + 2e^- + h\nu \rightarrow H_2 \quad (E_0 = 0.00\,V\text{ vs RHE})$$

The minimum bandgap of the semiconductor in photocatalysis of water splitting is 1.23 eV (~1100 nm). The bandgap of a photocatalyst should be less than 3.0 eV ($\lambda > 415$ nm). The band positions of different semiconductors for water splitting are shown in Figure 8.10. Mostly, semiconductors such as ZrO_2, $KTaO_3$, $SrTiO_3$ and TiO_2 have suitable bandgap in order to split water photochemically. However, carbon-based materials upon doping with other elements and molecules show superior and

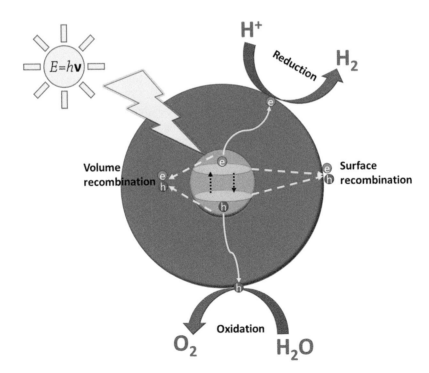

FIGURE 8.9 The mechanism of photocatalysis of water splitting.

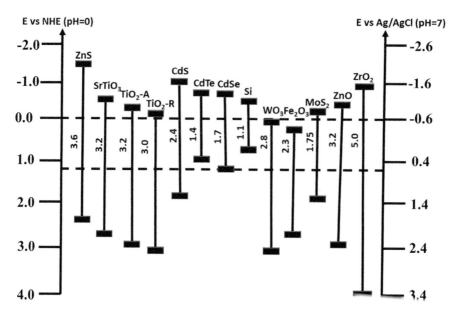

FIGURE 8.10 Bandgap positions of various semiconducting materials in photocatalysis at varying pH levels.

efficient performance by tuning the structure as well as the electronic properties of the carbon-based catalyst.

Currently, graphene and its derivatives show enormous potential in photocatalysis. The formation of a graphene/graphene oxide heterojunction with metal doping has been investigated to achieve high H_2 production [35]. Core-shell structures have proven to perform the best photocatalytic activity due to the synergistic effect between the core and the shell. Basically, the formation of the p-n junction between the inner core and outer shell ease in the enhancement of the charge carrier concentration onto the surface of the catalyst. Akansha and co-workers have reported carbon quantum dots (CQDs) on gold (Au) – that is Au@CQDs and CQDs photoelectrodes – that exhibited higher current density and superior hydrogen evolution rate [36]. The coating of the semiconductor with graphene and graphene-derived materials provides better performance for water splitting due to the availability of charge carriers on the surface of the graphene sheet [37]. The doping into the sheets of graphene can be introduced by lowering the thickness of the sheet that results in the opening of the bandgap due to quantum confinement. The presence of defects in the sheet of graphene also results in the opening of the bandgap. The vacancies and dislocations also open the bandgap by inducing additional electronic states. The involvement of some elements like F, H, O, N, etc. affects the electronic states of the sheets, for example fluorographene (2D sheet of graphene and fluoride) which is an insulator whose bandgap can be tailored by changing the concentration of fluoride. Li et al. demonstrated water splitting through a photocatalysis method taking graphene sheets

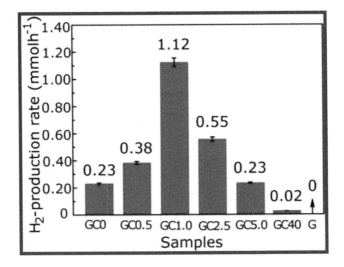

FIGURE 8.11 Comparison of photocatalytic activity at different %wt of graphene sheet decorated with cadmium sulphide (GC). (Reproduced with permission from Q. Li et al., *J. Am. Chem. Soc.* 133, 2011, 10878–10884. doi: 10.1021/ja2025454)

decorated with cadmium sulphide (CdS) clusters (Figure 8.11). Theoretical studies have shown that N-doped and P-doped graphene have lower valance bands and are the best selection of HER applications.

8.2.3 Photoelectrochemical Water Splitting

For solar-to-fuel conversion, PEC is a prominent method that mimics the natural photosynthesis process. PEC was first demonstrated in 1972 by Fujishima and Honda using titania as an anode and platinum as a cathode. Since then, tremendous research and development have been practised to split water photoelectrochemically basically in the field of catalyst design [38]. Few visible light-responsive oxide (WO_3, Fe_2O_3 and $BiVO_4$) catalysts are incapable of reducing H_2 because their conduction bands are low, and to overcome this limitation associated with these catalysts, an external bias on both the electrodes is applied in order to split the water. In PEC water splitting, water splits with the help of electrical charges (electrons and holes) that were generated when the light irradiates the catalyst. Since the thermodynamic potential of water splitting is 1.23 V, a voltage of approximately 2 V must be applied in order to split the water. For a material to split water, its bandgap should be larger than 1.23 eV, and in order to compensate for various losses, the bandgap must be between 1.5 and 3 eV (between ~800 and 500 nm) [39]. The catalyst can be n-type or p-type in which holes or electrons are generated after absorbing the light. For example, for an n-type semiconductor, electrons move towards the cathode to reduce water to H_2, and holes at the anode oxidise water to produce O_2 (Figure 8.12).

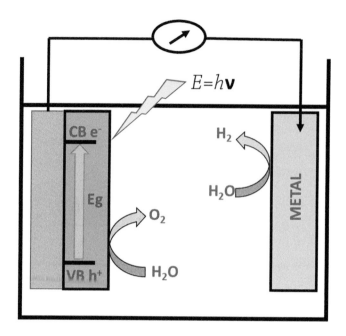

FIGURE 8.12 Photoelectrocatalysis in water splitting.

8.3 CARBON DIOXIDE REDUCTION

Carbon dioxide (CO_2) is a greenhouse gas that is released into the atmosphere from the burning of fossil fuels, leading the world towards serious global warming. Knowing the immensity of the problem, over the past few decades the reduction of CO_2 is gaining much attention as well as investment at the research and industrial levels. Reducing CO_2 levels and further regenerating CO_2 into eco-friendly fuels and chemicals is an efficient way to tackle this major problem. Talking about the different strategies of CO_2 reduction, the electrochemical and photochemical approaches are very efficient and attractive because the required energy as an input can be supplied from renewable energy sources, for example solar energy. The scheme for CO_2 reduction, its cycle and reduction potentials are shown in Figure 8.13. Royer first reported the reduction of CO_2 to formic acid on zinc electrodes in 1870 [40]. CO_2 reduction is a chemical process where carbon dioxide is reduced to carbon monoxide or carbonaceous fuel and chemicals by the means of some source of energy. When the energy to reduce CO_2 is electric current, that process is called *electrocatalytic* reduction of CO_2, and when the source of energy is light then the reduction process is called *photocatalytic* CO_2 reduction. In the photoelectrochemical process, both the light as well as electricity are utilised simultaneously to reduce CO_2. The current research aims to develop such a catalyst that reduces CO_2 in an efficient way (low overpotential and high selectivity) and for which the regenerated carbonaceous by-product does not compromise purity. There are basically two routes to reduce CO_2; one route utilises photovoltaics (PVs) to give sufficient voltage at the cathode for reduction and anode for the oxidation of water. Here, this route has an advantage

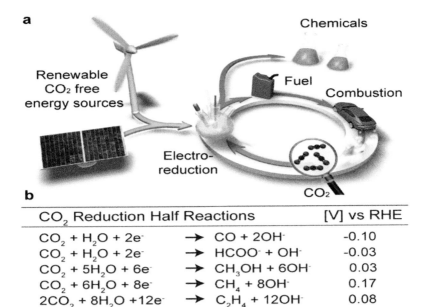

FIGURE 8.13 (a) An artificial method of carbon recycling powered by renewable electricity produced from wind and solar sources. (b) Reduction potentials of CO_2 reduction reactions along with H_2 evolution reaction. (Reproduced with permission from K.P. Kuhl et al., *J. Am. Chem. Soc.* 136, 2014, 14107–14113. doi: 10.1021/ja505791r.)

in that both the PV and electrocatalyst can be utilised simultaneously (called the photoelectrochemical method) to gain the combined performance of both units. The second route is the use of direct light, where light-absorbing and light-activated semiconductors are used as a cathode. The system is very simple as the design is wireless and compact. The future of CO_2 reduction strongly depends upon the design and modulation of electrocatalyst and photocatalyst.

CO_2 reduction generates different products including carbon monoxides, formic acids, methanol, methane, etc. Since CO_2 produces some chemicals as a by-product while reducing it, among the resultant chemicals, ethanol and hydrocarbons are attractive because of their potential applications and economic importance. The generation of each product has a thermodynamic reduction potential at standard temperature and pressure (STP) in 1M of other solute at pH 7 and potential (V) versus RHE:

$$CO_2(g) + e^- \rightarrow {}^*COO^- \qquad E^\circ = -1.90\,V$$

$$CO_2(g) + 2H^+ + 2e^- \rightarrow HCOOH(l) \qquad E^\circ = -0.61\,V$$

$$CO_2(g) + H_2O(l) + 2e^- \rightarrow HCOO^-(aq) + OH^- \qquad E^\circ = -0.43\,V$$

$$CO_2(g) + 2H^+ + 2e^- \rightarrow CO(g) + H_2O \qquad E^\circ = -0.53\,V$$

$$CO_2(g) + H_2O(l) + 2e^- \rightarrow CO(g) + 2OH^- \qquad E^\circ = -0.52\,V$$

$$CO_2(g) + 4H^+ + 2e^- \rightarrow HCHO(l) + H_2O(l) \qquad E^\circ = -0.48\,V$$

$$CO_2(g) + 3H_2O(l) + 4e^- \rightarrow HCHO(l) + 4OH^- \qquad E^\circ = -0.89\,V$$

$$CO_2(g) + 6H^+ + 6e^- \rightarrow CH_3OH(l) + H_2O(l) \qquad E^\circ = -0.38\,V$$

$$CO_2(g) + 5H_2O(l) + 6e^- \rightarrow CH_3OH(l) + 6OH^- \qquad E^\circ = -0.81\,V$$

$$CO_2(g) + 8H^+ + 8e^- \rightarrow CH_4(g) + 2H_2O(l) \qquad E^\circ = -0.24\,V$$

$$CO_2(g) + 6H_2O(l) + 8e^- \rightarrow CH_4(g) + 8OH^- \qquad E^\circ = -0.25\,V$$

$$2CO_2(g) + 12H^+ + 12e^- \rightarrow C_2H_4(g) + 4H_2O(l) \qquad E^\circ = -0.06\,V$$

$$2CO_2(g) + 8H_2O(l) + 12e^- \rightarrow C_2H_4(g) + 12OH^- \qquad E^\circ = -0.34\,V$$

$$2CO_2(g) + 12H^+ + 12e^- \rightarrow CH_3CH_2OH(l) + 3H_2O(l) \qquad E^\circ = -0.08\,V$$

$$2CO_2(g) + 9H_2O(l) + 12e^- \rightarrow CH_3CH_2OH(l) + 12OH^- \qquad E^\circ = -0.33\,V$$

8.3.1 Electrochemical CO_2 Reduction

Electrochemical reduction of CO_2 is a multi-step reaction that generally involves two, four, six, eight or twelve electrons for the reduction that takes place at the boundary of an electrode-electrolyte interface. The electrolyte is CO_2 saturated. A schematic representation of CO_2 reduction is shown in Figure 8.14. The steps involved in the reduction are as follows:

1. CO_2 adsorbs on the surface of an electrocatalyst which is a rate-determining step. This step requires large amounts of energy and an overpotential of $-1.9\,V$ versus RHE is needed in order to continue the sluggish reaction.
2. The bonds between the C and O in CO_2 start transforming to C-H bonds due to the transfer of electrons. This bond transformation is due to CO_2^- radical which is very reactive and undergoes several electron and proton transformations to yield the products.
3. Diffusion of the regenerated products into the electrolyte and desorption from the surface of the electrocatalyst take place.

The final product obtained in the reduction process depends upon the type of catalysts used and the potential applied. Research has proven that several metals such as Cu, Au,

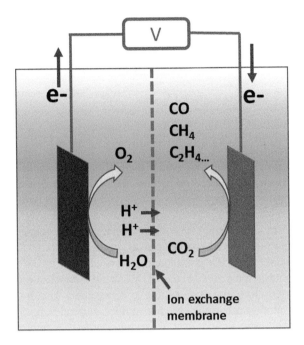

FIGURE 8.14 Electrochemical CO_2 reduction scheme.

In, Sn, Pb, Zn, Ag, Pd, Bi and carbon-based materials are catalytically good candidates [41–45]. Elements of group 1 metals such as Sn, Hg, Pb, In, etc. produce formic acid as their main regenerated product [44]. CO is the main product in CO_2 electroreduction when the catalysts are the metals, such as Au, Ag, Zn and Pd [46,47]. Copper from the group 3 family is the only electrocatalyst that converts CO_2 into hydrocarbons and oxygenates. Therefore, copper is gaining much attention as an electrocatalyst for CO_2 reduction in order to reduce the overpotential and increase efficiency. There are various parameters used for the analysis of catalytic activity for CO_2 reduction such as reduction efficiency (faradic efficiency), overpotential, current density, etc.

The efficiency of CO_2 reduction is basically calculated using two methods: faradic efficiency (FE) and energy efficiency (EF).

Faradic efficiency is calculated as the percentage of electrons that are utilised in the regeneration of a resultant product. The expression for faradic efficiency is

$$E_{faradic} = \alpha nF/Q$$

where α is the number of electrons transferred, n is the number of moles of the target product, F is the faradic constant and Q is the amount of charge consumed or passed.

Energy efficiency is the ratio of the amount of energy received into the product and the amount of input electrical energy to drive the reaction. This can be expressed as

$$E_{energy} = \frac{E_{eq}}{E_{eq}\eta} \times E_{faradic}$$

where E_{eq} is the equilibrium potential, η is the overpotential and $E_{faradic}$ is the faradic efficiency.

The overpotential is the variation in the potential from experimental value and thermodynamically obtained value. It is given as

$$\eta = (E_{applied} - E_o)$$

where E_o is the standard reduction potential, and $E_{applied}$ is the experimentally determined potential.

When the magnitude current obtained is divided by the geometric surface area of the working electrode, then that is called the *current density*. It is a very important parameter in catalysis, as the performance of the catalyst largely depends upon the surface area of the catalyst.

Carbonaceous materials may be very promising in the reduction of CO_2, as carbon-based materials possess good electrical conductivity, large surface area, low cost and excellent stability. But pristine carbon-based materials are inert in nature and show very poor activity in CO_2 reduction. So, to compensate for the inertness of the carbon-based electrocatalyst, doping is being done. The doping is basically done using heteroatoms such as nitrides, phosphides, sulphides, borides, etc. in order to introduce defects into the catalyst. Currently, various forms of carbonaceous materials such as graphene sheets and quantum dots, CNTs, carbon nanofibers, etc. are being doped and tailored with heteroatoms and different elements.

The active sites of the carbon atom were stated to be highly positive when doped with electronegative nitrogen (N dopant). Salehi et al. synthesised N-doped carbon nanofibers (CNFs) from electrospun polyacrylonitrile precursors which have shown a small onset potential of only 170 mV for reducing CO_2 to CO [48]. Einaga and co-workers successfully prepared sp3-bonded carbons along boron-doping diamonds that have shown a faradic efficiency of 74% in methanol as well as in seawater [9]. Ajayan and co-workers N-doped graphene quantum dots and reported that CO_2 can be reduced into hydrocarbons and oxygenates with a high faradic efficiency of 90%, high selectivity of ethanol and ethylene reaching 45% [49]. Quan et al. demonstrated that N-doped nanodiamond shows a faradic efficiency of 77% for the production of acetone and 15% efficiency towards the formation of formate [50]. The CO_2 reduction activity is shown in Figure 8.15 using cyclic voltammetry [58].

8.3.2 PHOTOCHEMICAL CO_2 REDUCTION

Photochemical CO_2 conversion consists of a multi-step process:

1. Light absorption to generate electron-hole pair
2. Charge (e^-/h^+) separation
3. CO_2 adsorption on the surface of the catalyst
4. Surface reactions
5. Desorption of regenerated product into the solution

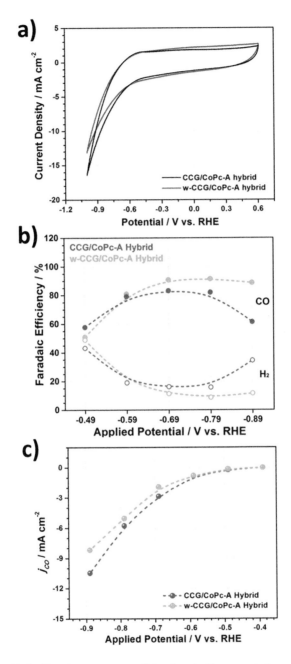

FIGURE 8.15 (a) Cyclic voltammograms of the chemically converted graphene and Co(II) phthalocyanine hybrid (CCG/CoPc-A) and washed-CCG/CoPc-A hybrids in CO_2-saturated 0.1 M $KHCO_3$ electrolyte. (b) The result of gas analysis of the w-CCG/CoPc-A and CCG/CoPc-A hybrid catalysts at different overpotentials. (c) CO partial current density of the w-CCG/CoPc-A and CCG/CoPc-A hybrid catalysts. (Reproduced with permission from J. Choi et al., *ACS Energy Lett.* 4(3), 2019, 666–672.)

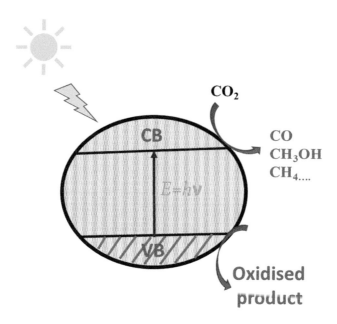

FIGURE 8.16 Electrochemical CO_2 reduction scheme.

When the light is illuminated onto the surface of the photocatalyst, then the electron moves from valance band (VB) to conduction band (CB) leaving behind an equal number of holes in VB. Band structure of the catalyst plays a crucial role in reducing CO_2 and oxidising water (Figure 8.16). The band.gap should be sufficient enough to overcome the extra overpotential associated with the reactions and the systems. But the bandgap must not be very large because the large bandgap limits solar power utilisation. The CB edge should be more negative than the redox potential of reduction of CO_2, and the VB must be more positive than the oxidation potential of water.

Carbonaceous materials are garnering attention in the field of photocatalysis of CO_2 reduction. Basically, graphitic carbon nitride (g-C_3N_4) is at its apex because of its ideal band positions and layered structure. In 2009, Domen and co-workers reported its photocatalytic activity for the first time for hydrogen generation in the presence of visible light [51]. Since then, various efforts have been made to investigate the catalytic activity of C_3N_4. Peng et al. studied the photocatalytic behaviour of C_3N_4 and reported that porous C_3N_4 prepared from urea yields methanol and ethanol as the regenerated products, and non-porous C_3N_4 derived from melamine only gives ethanol [52]. The exfoliation of bulk C_3N_4 powder to a thickness of a single atom increases the surface area and charge separation and ultimately boosts the photocatalytic activity [53]. Ye et al. demonstrated that upon hybridising, exfoliated C_3N_4 nanosheet with zirconia (Zr)-derived MOF showed better charge separation efficiency of the catalyst and enhanced carrier lifetime [54]. Graphene oxide (GO) can also reduce CO_2 due to its suitably negative conduction band (0.79 V versus RHE) and positive valence

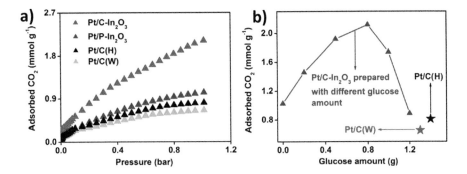

FIGURE 8.17 (a) CO_2 adsorption capacities of platinum co-catalyst/carbon-coated In_2O_3(Pt/C- In_2O_3), pure In_2O_3(Pt/-P-In_2O_3), hydrothermally synthesised (Pt/C[H]), and washed (Pt/C[W]). (b) CO_2 adsorption capacity with the variation of glucose amount of the catalyst. Black and red stars represent the CO_2 adsorption capacities of Pt/C(H) and Pt/C(W), respectively. (Reproduced with permission from Y. Pan et al., *J. Am. Chem. Soc.* 139, 2017, 4123–4129. doi: 10.1021/jacs.7b00266.)

band (+2.91 V versus RHE) to enable water oxidation for the generation of desired product [55].

Different carbon-based material coatings are also performed on different materials to exploit the extraordinary properties of carbon in enhancing the catalytic activities. For example, Pan et al. have done a coating of 5 nm-thick carbon layer on indium-oxide (In_2O_3) nanobelts that showed a satisfactory photocatalytic reduction of CO_2 to CO and CH_4 because of the fact that the carbon coating improves the absorption capacity of the catalyst (Figure 8.17) that boosts the rate of electron-hole pair generation and separation [56].

8.3.3 Photoelectrochemical CO_2 Reduction

The PEC method for the reduction of CO_2 has been a matter of interest because of its efficient results and performance as compared to photochemical and electrochemical systems. The photoelectrode in a photoelectrochemical system utilises light to generate electron-hole pairs at one-half of the cell, while the other half of the reaction takes place at the counter-electrode with the help of external bias (Figure 8.18). Here light and potential are used simultaneously. Proton exchange membranes are used here to separate oxidation and reduction products, and the regenerated product can be collected at the end. Hence, CO_2 reduction by this technique is regarded as one of the best routes to gain better catalytic activity. However, the reduction of H_2O at the other half of the reaction hinders the conversion efficiency and faradic efficiency. The selection of material is also one of the reasons for degrading the conversion efficiency, because improper bandgap selection of catalyst material restricts the sufficient amount of light to trigger the charge carriers present in the semiconductor which eventually increases the overpotential.

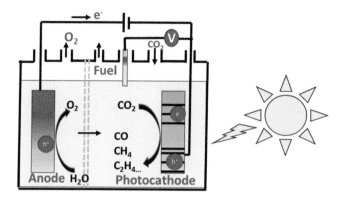

FIGURE 8.18 Photoelectrochemical CO_2 reduction scheme.

8.4 CONCLUSION

In this chapter we discussed the techniques of energy generation, conversion and its practices. A clean and green approach for the generation as well as conversion of energy is possible if we generate and convert it in a sustainable way using renewable resources. The generation of H_2 gas as a fuel through the splitting of water is the best practice to eliminate conventional energy sources that can harm the earth and environment by generating CO_2 gas that leads the earth towards global warming. The reduction of carbon dioxide that is present in the atmosphere due to the burning of conventional fuels is an essential goal. Through electrocatalysis, photocatalysis and photoelectrocatalysis, these serious issues can be addressed efficiently and in a sustainable way. Carbon is the best choice as an electrocatalyst and photocatalyst because it is renewable, inexpensive and earth abundant. The hope is to create a clean and green earth where all survive in harmony and lead healthy lives in a healthy nature and environment.

REFERENCES

1. M. Notarianni, J. Liu, K. Vernon, N. Motta, Synthesis and applications of carbon nanomaterials for energy generation and storage, *Beilstein J. Nanotechnol.* 7, 2016, 149–196. doi: 10.3762/bjnano.7.17
2. A.R. Zeradjanin, J. Grote, G. Polymeros, K.J.J. Mayrhofer, A critical review on hydrogen evolution electrocatalysis: Re-exploring the volcano-relationship, *Electroanalysis* 28, 2016, 2256–2269. doi: 10.1002/elan.201600270
3. Y. Cai, C. Ma, Y. Zhu, J.X. Wang, R.R. Adzic, Low-coordination sites in oxygen-reduction electrocatalysis: Their roles and methods for removal, *Langmuir* 27, 2011, 8540–8547, doi.org/10.1021/la200753z
4. J. Xing, Q. Fang, J. Zhao, H. Gui, Inorganic photocatalysts for overall water splitting, *Chem. Asian J.* 7, 2012, 642–657. doi: 10.1002/asia.201100772
5. Y. Xu, R. Xu, Metal-free carbonaceous electrocatalysts and photocatalysts for water splitting, *Chem. Soc. Rev.* 45, 2016, 3039–3052. doi: 10.1039/C5CS00729A
6. J. Zhang, G. Chen, K. Müllen, X. Feng, Carbon-rich nanomaterials: Fascinating hydrogen and oxygen electrocatalysts, *Adv Mater*, 1800528, 2018, 1–22. doi: 10.1002/adma.201800528

7. K. Dai, T. Peng, K. Noda, M. Hattori, K. Amari, K. Kakehi, S. Noda, Photocatalytic Hydrogen Production on Nanocomposite of Carbon Nanotubes and TiO_2 Photocatalytic Hydrogen Production on Nanocomposite of Carbon Nanotubes and TiO_2, *J. Phys. Conf. Ser.*, 1032, 2018, 012056. doi: 10.1088/1742-6596/1032/1/012056
8. X. Wang, M. Liu, Q. Chen, K. Zhang, J. Chen, M. Wang, P. Guo, L. Guo, Synthesis of CdS/CNTs photocatalysts and study of hydrogen production by photocatalytic water splitting, *Int. J. Hydrogen Energy*, 38, 2013, 6–11. doi: 10.1016/j.ijhydene.2013.03.016
9. A. Ali, D. Akyüz, M. Adeel Asghar, A. Koca, B. Keskin, Free-standing carbon nanotubes as non-metal electrocatalyst for oxygen evolution reaction in water splitting, *Int. J. Hydrogen Energy*, 43, 2018, 1123–1128. doi: 10.1016/j.ijhydene.2017.11.060
10. S. Ye, R. Wang, M. Wu, Y. Yuan, A review on g-C_3N_4 for photocatalytic water splitting and CO_2 reduction, *Appl. Surf. Sci.* 358, 2015, 15–27. doi: 10.1016/j.apsusc.2015.08.173
11. Q. Xiang, B. Cheng, J. Yu, Graphene-based photocatalysts for solar-fuel generation, *Angew. Chem. Int. Ed. Engl.* 54, 2015, 11350–11366. doi: 10.1002/anie.201411096
12. S. Patnaik, S. Martha, K.M. Parida, RSC Advances photocatalytic hydrogen production, *RSC Adv.* 6, 2016, 46929–46951. doi: 10.1039/C5RA26702A
13. S.X. Zou, Y. Zhang, As featured in:, Noble metal-free hydrogen evolution catalysts for water splitting, *Chem. Soc. Rev.* 44, 2015, 5148–5180. doi: 10.1039/c4cs00448e
14. D. Strmcnik, P.P. Lopes, B. Genorio, V.R. Stamenkovic, N.M. Markovic, Design principles for hydrogen evolution reaction catalyst materials, *Nano Energy*, 29, 2016, 29–36. doi:10.1016/j.nanoen.2016.04.017.
15. R. Subbaraman, D. Tripkovic, K. Chang, D. Strmcnik, A.P. Paulikas, P. Hirunsit, M. Chan, J. Greeley, V. Stamenkovic, N.M. Markovic, Trends in activity for the water electrolyser reactions on 3d M(Ni,Co,Fe,Mn) hydr(oxy)oxide catalysts, *Nat. Mater.* 11, 2012, 550–557. doi: 10.1038/nmat3313
16. X. Li, X. Hao, A. Abudula, G. Guan, Nanostructured catalysts for electrochemical water splitting: Current state and prospects, *J. Mater. Chem. A.* 4, 2016, 11973–12000. doi: 10.1039/c6ta02334g
17. E. Fabbri, A. Habereder, K. Waltar, R. Kötz and T. J. Schmidt, Developments and perspectives of oxide-based catalysts for the oxygen evolution reaction *Catal. Sci. Technol.* 4, 2014, 3800–3821. doi: 10.1039/C4CY00669K
18. I.C. Man, H. Su, F. Calle-vallejo, H.A. Hansen, J.I. Martínez, N.G. Inoglu, J. Kitchin, T.F. Jaramillo, J.K. Nørskov, J. Rossmeisl, Universality in oxygen evolution electrocatalysis on oxide surfaces, *ChemCatChem* 3, 2011, 1159–1165. doi: 10.1002/cctc.201000397
19. A.J. Bard, L.R. Faulkner, E. Swain, C. Robey, Fundamentals and Applications, n.d.
20. Y. Gorlin, T.F. Jaramillo, A Bifunctional nonprecious metal catalyst for oxygen reduction and water oxidation, *J. Am. Chem. Soc.* 132, 2010, 13612–13614.
21. S. Mo, Building an appropriate active-site motif into a hydrogen-evolution catalyst with thiomolybdate $[Mo_3S_{13}]^{2-}$ clusters, *Nat. Chem.* 6, 2014, 248–253. doi: 10.1038/nchem.1853
22. S. Chen, J. Duan, M. Jaroniec, S.Z. Qiao, Three-dimensional N-doped graphene hydrogel/NiCo double hydroxide electrocatalysts for highly efficient oxygen evolution, *Angew. Chemie, Int. Ed.* 52, 2013, 13567–13570. doi: 10.1002/anie.201306166
23. J.M. Rubi and S. Kjelstrup, Mesoscopic nonequilibrium thermodynamics gives the same thermodynamic basis to butler–volmer and nernst equations. *J. Phys. Chem. B.* 107, 2003, 13471–13477. doi: 10.1021/jp030572g
24. C. Hu, D. Liu, Y. Xiao, L. Dai, Functionalization of graphene materials by heteroatom-doping for energy conversion and storage, *Prog. Nat. Sci. Mater. Int.* 28, 2018, 121–132. doi: 10.1016/j.pnsc.2018.02.001
25. Y. Yan, X. Ge, Z. Liu, J. Wang, J. Lee, X. Wang, V.A. Online, Facile synthesis of low crystalline MoS2 nanosheet-coated CNTs for enhanced hydrogen evolution reaction, *Nanoscale*, 5, 2013, 7768–7771. doi: 10.1039/c3nr02994h

26. J. Deng, W. Yuan, P. Ren, Y. Wang, D. Deng, Z. Zhang, X. Bao, High-performance hydrogen evolution electrocatalysis by layer-controlled MoS_2 nanosheets, *RSC Adv.* 4, 2014, 34733–34738. doi: 10.1039/C4RA05614K
27. Han Zhu, MingLiang Du, Ming Zhang, MeiLing Zou, TingTing Yang, YaQin Fu, JuMing Yao, The design and construction of 3D rose-petal-shaped MoS_2 hierarchical nanostructures with structure-sensitive properties, *J. Mater. Chem. A*, 2, 2014, 7680–7685 doi: 10.1039/c4ta01004c
28. D.H. Youn, S. Han, J.Y. Kim, J.Y. Kim, H. Park, S.H. Choi, J.S. Lee, Highly active and stable hydrogen evolution electrocatalysts based on molybdenum compounds on carbon nanotube à graphene hybrid support, *ACS Nano*, 8, 2014, 5164–5173. doi: 10.1021/nn5012144
29. L. Liao, J. Zhu, X. Bian, L. Zhu, M.D. Scanlon, H.H. Girault, MoS_2 formed on mesoporous graphene as a highly active catalyst for hydrogen evolution, *Adv. Funct. Mater.* 23, 2013, 5326–5333. doi: 10.1002/adfm.201300318
30. J. Liu, Y. Liu, N. Liu, Y. Han, X. Zhang, H. Huang, Y. Lifshitz, S. Lee, J. Zhong, Z. Kang, Metal-free efficient photocatalyst for stable visible water splitting via a two-electron pathway, 347, 2015, 1–6.
31. Xinyuan Xia, Ning Deng, Guanwei Cui, Junfeng Xie, Xifeng Shi, Yingqiang Zhao, Qian Wang, Wen Wang, Bo Tang, NIR light induced H_2 evolution by a metal-free photocatalyst, *Chem. Commun.*, 51, 2015, 10899–10902, doi:10.1039/C5CC02589C
32. J. Zhang, C. Zhang, J. Sha, H. Fei, Y. Li, J.M. Tour, Efficient water splitting electrodes based on laser-induced graphene, *ACS Appl Mater Interfaces*, 9, 2017. doi: 10.1021/acsami.7b06727
33. Y. Chen, X. Zhang, Z. Zhou, Carbon-based substrates for highly dispersed nanoparticle and even single-atom electrocatalysts, 1900050, 2019, 1–9. doi: 10.1002/smtd.201900050
34. X. Wang, H. Chen, Y. Xu, J. Liao, B. Chen, H-S. Rao, D-B. Kuang, C-Y. Su, Self-supported $NiMoP_2$ nanowires on carbon cloth as an efficient and durable electrocatalyst for overall water splitting, *J. Mater. Chem. A*, 5, 2017, 7191–7199. doi: 10.1039/c6ta11188b
35. T. Yeh, C. Teng, S. Chen, H. Teng, Nitrogen-doped graphene oxide quantum dots as photocatalysts for overall water-splitting under visible light illumination, *Adv Mater*, 26, 2014, 3297–3303. doi: 10.1002/adma.201305299
36. Akansha Mehta, Pooja D., Anupma Thakur, Soumen Basu, Enhanced photocatalytic water splitting by gold carbon dot core shell nanocatalyst under visible/sunlight, *New J. Chem.* 41, 2017, 4573–4581. doi: 10.1039/C7NJ00933J
37. Zheng Wang, Jungang Hou, Chao Yang, Shuqiang Jiao and Hongmin Zhu, Three-dimensional MoS_2–CdS–c-TaON hollow composites for enhanced visible-light-driven hydrogen evolution, *Chem. Commun.* 2014, 50, 1731–1734. doi: 10.1039/c3cc48752k
38. E.L. Miller, Photoelectrochemical water splitting, *Energy Environ. Sci.*, 8, 2015, 2809–2810. doi: 10.1039/C5EE90047F
39. M.G. Walter, E.L. Warren, J.R. Mckone, S.W. Boettcher, Q. Mi, E.A. Santori, N.S. Lewis, Solar water splitting cells, *Chem. Rev.* 110, 2010, 6446–6473. doi: 10.1021/cr1002326
40. D.P. Summers, S. Leach, K.W. Frese Jr., The electrochemical reduction of aqueous carbon dioxide to methanol at molybdenum electrodes with low overpotentials, *J. Electroanal. Chem.* 205, 1986, 219–232. doi: 10.1016/0022-0728(86)90233-0.
41. C.S. Chen, A.D. Handoko, J.H. Wan, L. Ma, D. Ren, B.S. Yeo, Stable and selective electrochemical reduction of carbon dioxide to ethylene on copper mesocrystals, *Catal. Sci. Technol.* 2015, 161–168. doi: 10.1039/c4cy00906a
42. S. Zhang, P. Kang, S. Ubnoske, M.K. Brennaman, N. Song, R.L. House, T. Glass, T.J. Meyer, Polyethylenimine-enhanced electrocatalytic reduction of CO_2 to formate at nitrogen-doped carbon nanomaterials, *J. Am. Chem. Soc.* 136, 2014, 7845–7848. doi: 10.1021/ja5031529

43. N. Hoshi, M. Kato, Y. Hori, Electrochemical reduction of CO, on single crystal electrodes of silver Ag(111), Ag(100) and Ag(110), *J. Electroanalytical Chem.* 440, 1997, 283–286, doi.org/10.1016/S0022-0728(97)00447-6
44. Y. Hori, H.H.I. Wakebe, T. Tsukamoto, O. Koga, Electrocatalytic process of CO selectivity in electrochemical reduction of CO_2 at metal electrodes in aqueous media, *Electrochim. Acta* 39, 1994, doi.org/10.1016/0013-4686(94)85172-7
45. J.L. Dimeglio, J. Rosenthal, Selective conversion of CO_2 to CO with high efficiency using an inexpensive bismuth-based electrocatalyst, *J. Am. Chem. Soc.* 135, 2013, 8798–8801. doi: 10.1021/ja4033549
46. B. Aurian-Blajeni, Electrochemical reduction of carbon dioxide, in: *Electrochemistry in Transition*, Conway, B.E., Murphy, O.J., Srinivasan, S. (Eds.), 1992, pp. 381–396, Springer.
47. C. M. Sánchez-Sánchez, V. Montiel, D.A. Tryk, A. Aldaz, A. Fujishima, Electrochemical approaches to alleviation of the problem of carbon dioxide accumulation, *Pure Appl. Chem.*,73, 2001, 1917–1927. doi: 10.1351/pac200173121917.
48. B. Kumar, M. Asadi, D. Pisasale, S. Sinha-ray, B.A. Rosen, R. Haasch, J. Abiade, A.L. Yarin, A. Salehi-khojin, Catalysts for carbon dioxide reduction, *Nat. Commun.* 4, 2013, 1–8. doi: 10.1038/ncomms3819
49. J. Wu, S. Ma, J. Sun, J.I. Gold, C. Tiwary, B. Kim, L. et al., Oxygenates, *Nat. Commun.* 7, 2016, 1–6. doi: 10.1038/ncomms13869
50. Y. Chen, C.W. Li, M.W. Kanan, Aqueous CO_2 reduction at very low overpotential on oxide-derived Au nanoparticles, *J. Am. Chem. Soc.* 4, 2012, 1–4.
51. X. Wang, K. Maeda, A. Thomas, K. Takanabe, G. Xin, J.M. Carlsson, K. Domen, M. Antonietti, A metal-free polymeric photocatalyst for hydrogen production from water under visible light, *Nat. Mater.* 8, 2008, 76–80. doi: 10.1038/nmat2317
52. J. Mao, T. Peng, X. Zhang, K. Li, L. Ye, L. Zan, Effect of graphitic carbon nitride microstructures on the activity and selectivity of photocatalytic CO_2 reduction under visible light, *Catal. Sci. Technol.* 3, 2013, 1253–1260. doi: 10.1039/c3cy20822b
53. S. Yang, Y. Gong, J. Zhang, L. Zhan, L. Ma, Z. Fang, X. Wang, P.M. Ajayan, Exfoliated graphitic carbon nitride nanosheets as efficient catalysts for hydrogen evolution under visible light, *Adv. Mater.* 25, 2013, 2452–2456. doi: 10.1002/adma.201204453.
54. L. Shi, T. Wang, H. Zhang, K. Chang, J. Ye, Electrostatic self-assembly of nanosized carbon nitride nanosheet onto a zirconium metal – Organic framework for enhanced photocatalytic CO_2 reduction, *Adv. Funct. Mater.* 25, 2015, 5360–5367. doi: 10.1002/adfm.201502253
55. H-C. Hsu, I. Shown, H-Y. Wei, Y-C. Chang, H-Y. Du, Y-G. Lin, C-A. Tseng, C-H. Wang, L-C. Chen, Y-C. Lin, K-H. Chen, Graphene oxide as a promising photocatalyst for CO_2 to methanol conversion, *Nanoscale* 5, 2013, 262–268. doi: 10.1039/c2nr31718d
56. Y. Pan, Y. You, S. Xin, Y. Li, G. Fu, Z. Cui, Y. Men, F. Cao, S. Yu, J.B. Goodenough, Photocatalytic CO_2 reduction by carbon-coated indium-oxide nanobelts, *J. Am. Chem. Soc.* 139, 2017, 4123–4129. doi: 10.1021/jacs.7b00266
57. K.P. Kuhl, T. Hatsukade, E.R. Cave, D.N. Abram, J. Kibsgaard, T.F. Jaramillo, Electrocatalytic conversion of carbon dioxide to methane and methanol on transition metal surfaces, *J. Am. Chem. Soc.* 136, 2014, 14107–14113. doi: 10.1021/ja505791r.
58. J. Choi, P. Wagner, S. Gambhir, R. Jalili, D.R. MacFarlane, G.G. Wallace, D.L. Officer, Steric modification of a cobalt phthalocyanine/graphene catalyst to give enhanced and stable electrochemical CO_2 reduction to CO, *ACS Energy Lett.* 4, 2019, 666–672. doi: 10.1021/acsenergylett.8b02355
59. Q. Li, B. Guo, J. Yu, J. Ran, B. Zhang, H. Yan, J.R. Gong, Highly efficient visible-light-driven photocatalytic hydrogen production of CdS-cluster-decorated graphene nanosheets, *J. Am. Chem. Soc.* 133, 2011, 10878–10884. doi: 10.1021/ja2025454

9 Nanogenerator

9.1 INTRODUCTION

The day-by-day depletion of fossil fuels with increasing hazardous emissions has become a very serious issue for environmental conditions [1,2], leading to global warming and environmental pollution. So, it is one of the most urgent challenges to search for renewable and green energy resources for the development of human civilisation. We have to look forward for alternative large-scale energy sources rather than relying on worldwide well-known resources such as coal, petroleum, natural gases and nuclear. To power a city and even a country, macro-/mega-scale energy is required (Figure 9.1) [3]. Active research and development are continuously going on in the domain of researching for alternative energy resources such as solar, wind, geothermal, biological and hydrogen, etc. With the enduring growth of the internet of things (IOTs), small-scale energy production requires a suitable technology for independent, maintenance-free, long-lasting and continuous operation of portable and implantable biosensors to monitor health issues [4], nanorobotics [5], actuators [6], MEMS (micro-electromechanical systems) [7], wireless transmitters for security purposes, remote and mobile sensors and wearable electronic devices [8,9]. Nano-scale electronics operate at ultralow power consumption to a modest level (microwatt [μW] to milliwatt [mW]) which also need independent and continuous operational power sources [10]. Batteries are commonly used to power microelectronics, which gives rise to possible consequences such as limited lifetime, recycling, replacement of batteries, larger size as compared to nano devices and so on. Leakage and waste hazard chemicals in batteries will cause many environmental issues [11]. Therefore, self-powered nano devices without batteries are highly desired which can be helpful to reduce the size and total weight of the system as well as the adaptability of the system.

New technologies involve an emerging field of nano energy which can be harvested from the working environment and human activities to directly power self-sufficient nano/micro devices. The state-of-the-art small world of technology is known as nanotechnology. The goal of nanotechnology is to develop self-sufficient power sources by scavenging various forms of energy from the environment. Our environment is abundant with varieties of energy such as solar energy, wind power, thermal energy, chemical energy and various forms of mechanical energy, etc., which can be scavenged in many self-powered energy systems [12]. Among these energy resources, mechanical energy has attracted intense interest and is considered to be an effective and promising approach to resolving the energy issue, as it can be extracted from our living environment including human body motion [13]. Which form of energy needs to be harvested depends upon the working environment of the device. If the device is used in light, solar energy is harvested. If the device is used in the dark, mechanical vibration energy can be harvested. Nano-scale energies were invented

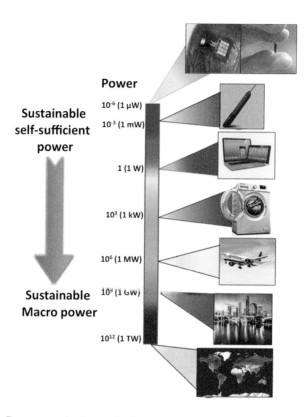

FIGURE 9.1 Power magnitude required to power the world, even a city, and all types of small electronics and bio-implanted devices.

in 2006 by Prof. Zhong Lin Wang and his group in ambient conditions through a nano device named a *nanogenerator* [2]. They converted mechanical energy into electrical energy by means of a piezoelectric zinc oxide nanowires (ZnONW) based nanogenerator.

9.1.1 Mechanical Energy Resources

Various resources of mechanical energy and their applications can be classified as follows:

- *Sporadic mechanical source*: Kinetic energy generated from human body parts through muscles, foot strike; biomechanical energy generated from motion of joints such as ankle, knee, hip, arm and elbow, fingers, blood flow, exhalation, etc. These sources can harvest power in the range of 60 µW cm^{-2} – 1.10 mW cm^{-2} (Figure 9.2) [14].
- *Mechanical energy*: This can be generated from human or animal activities such as walking and running.
- *Thermal energy*: Breathing out, exhalation process.

Nanogenerator

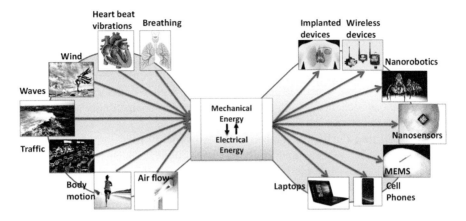

FIGURE 9.2 Harvesting energy resources and use of the energy in various applications.

- *Steady-state mechanical source*: Produces unlimited energy via wind, water flow, ocean waves and solar energy with an efficiency of 10%–24%; extracted energy is in the range of 100 µW cm^{-2} – 100 mW cm^{-2} (Figure 9.2) [14].
- *Mechanical vibration source*: Power machines, mechanical stress and strain from high-pressure motors, and waist rotations can be extracted and used as ambient mechanical energy sources. Power output is approximately 800 µW cm^{-3} (∼4 µW cm^{-2}) with an efficiency of 25%–50% [14].

9.1.2 Harvesting Mechanical Energy: Mechanism and Materials

A nano device used for the conversion of mechanical vibrations into electricity is termed a *nanogenerator*. The concept of a nanogenerator arises during the measurements of piezoelectric properties of ZnO nanowires (NW) by atomic force microscopy (AFM) technique [2]. The state-of-the-art three main transduction mechanisms such as electromagnetic effect [15–17], electrostatic effect [18] and piezoelectric effect [19–25] are used for vibration-to-electricity conversion. In 2014, the Zhang and Kim group designed an electromagnetic energy harvester based on the principle of Faraday's law of induction, which can convert vibrational energy into 263 mW power and lighten a candescent bulb [26]. Worldwide use of these techniques exhibits several challenges, such as dependence on external power sources [27], complex architecture [3] and need for high-quality materials [28]. Electromagnetic energy harvesters are limited to their bulky size and setup, and the power output signals rely purely on the number of coil turns and quality of magnets. Piezoelectric energy harvesters include only materials based on piezoelectric effects, such as inorganic materials like quartz crystal and ZnO nanowires, ceramics like lead zirconate titanate (PZT), barium titanate (BaTiO$_3$) and some polymers like poly(vinylidene fluoride) (PVDF), its co-polymer poly(vinylidenefluoride-*co*-tri-fluoroethylene) (P[VDF-TrFE]), some biological tissues and micro-organisms, etc. Some naturally occurring piezomaterials are dry bones, tendons, silk, woods, dentin and collagen. Initially, the piezoelectric transduction mechanism was invented by Pierre and Jacques Curie in 1880 [29]. A flexible piezoelectric energy conversion device was first demonstrated in 2009 by Choi et al. which includes one electrode with a nanorod array, grown on conductive indium tin oxide (ITO) coated polyethersulfone (PES) substrates and a second electrode with the same substrate coated with Pd-Au film [30]. Higher-output electrical signals were generated approximately 10 μA cm^{-2} under a compression of 0.9 kgf. The major challenge with this method is the non-compatibility to integrated circuit (IC) fabrications.

Among all transduction methods, electrostatic energy harvesters have become more promising due to their smaller size, high sensitivity, high energy density, high conversion efficiency and excellent compatibility towards ICs and MEMS processes [31]. It is feasible to fabricate a self-sufficient power pack by integrating an electrostatic micro power generator with wireless sensors.

In 2006, Sterken et al. proposed an electrostatic microgenerator with a power capability of 50 μW for a 0.2 cm^2 area [32]. Later, a significant amount of research was carried out in terms of electroactive tribo-materials to make the highly efficient triboelectric nanogenerators (TENG). Triboelectric nanogenerators works on two universally well-known effects: contact electrification and electrostatic induction. TENG is featured by its inherent characteristics such as extensive material source [33], easy and fast fabrication process [34] and robustness [35].

Various nanomaterials for nanogenerators have been reported in the literature which include all piezoelectric materials as well as tribomaterials. Among all piezoelectric materials, one-dimensional piezoelectric, wurtzite structured, semiconducting materials such as ZnO, CdS, GaN, etc. are intensively studied and explored in prototype devices. But these materials limit the practical applications due to their poor control of self-assembled nanostructures [22]. PZT and PVDF are alternatives to semiconducting materials with larger piezoelectric co-efficients. The fabrication and maintenance of these materials are relatively easy and cheap. The bottleneck of piezoelectric materials based nanogenerators towards the practical applications is relatively their short lifetimes and their use of metals as current collectors. To meet future applications such as touchscreens and wearable skin sensors, the replacement of metals in nanogenerators and a material with excellent opaque properties are required. To resolve the issues, considerable efforts have been made for metal-free current collectors and transparent active electrodes. Therefore, carbon-based materials such as carbon nanotubes (CNTs) and graphene films possessing outstanding mechanical and optical properties can be integrated with piezoelectric nanogenerators to produce transparent and excellent device performance.

9.2 CLASSIFICATION OF NANOGENERATORS

Nanogenerators can be classified into three main categories:

1. Piezoelectric nanogenerator (PENG)
2. Triboelectric nanogenerator (TENG)
3. Pyroelectric nanogenerator (PYNG)

9.2.1 PIEZOELECTRIC NANOGENERATORS

PENG is an energy harvesting device that is capable of converting kinetic energy into electricity using nano-structured piezoelectric materials. Piezoelectric materials gained great attention towards miniaturisation of devices, flexibility and sustained power sources.

9.2.1.1 Mechanism

The operational mechanism of PENG is based on the piezoelectric potential generated in a material by piezoelectric effect. Initially, Fermi levels of two electrodes applied on the piezoelectric material are electrostatically balanced. When external strain is applied on the electrodes, it produces piezopotential between the electrodes due to the polarisation charges and creates Fermi levels skewness at the contacts. Free charge carriers flow between the two electrodes externally to re-balance the asymmetry in Fermi levels electrostatically. The flow of free charge carriers produces an alternating current (AC) signal under periodic triggering by mechanical force. A sustained flow of alternating current with continuous finger tapping (mechanical force) will be a suitable and promising power source option for micro- or nanosystems.

PENG was first designed by using aligned ZnO nanowire arrays and Pt-coated Si probe tip of contact-mode AFM (Figure 9.3) [2]. Different mechanism is possible in piezoelectric Zinc Oxide (ZnO) nanowires. The applied force through the Pt-coated AFM tip in contact mode is perpendicular to the grown direction of nanowires, which deforms the nanowire. First, the AFM tip comes into contact with the elongated surface with positive potential of the nanowire. At the same time, a negative bias voltage at the Pt-ZnO interface appears and forms a reverse-biased Schottky diode with negligible current. The Schottky barrier at the Pt-ZnO interface on top of the nanowire is responsible for electricity generation. When the tip contacts the suppressed portion of the nanowire having negative piezopotential, the Pt-ZnO interface is positive-biased. A potential difference between the two sides gives rise to flow of a sharp peak current, that must again be re-balanced electrostatically

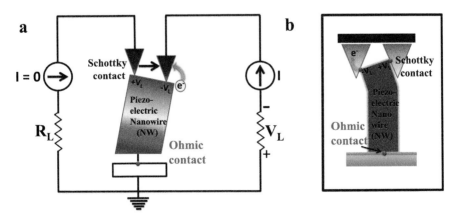

FIGURE 9.3 Force perpendicular to piezoelectric nanowire.

to nullify the potential difference. A single ZnO nanowire generates energy of 0.05 fJ, with a voltage output of 8 mV. It still demands advanced device design and materials due to the deformation in the nanowire with continuous strain applied by the AFM tip.

Zhong Lin Wang et al. further advanced their research. They designed a direct-current PENG using a ZnO nanowire array to harvest the high-frequency ultrasonic waves [36]. The operational mechanism is the same as explained earlier. Instead of using a Pt-coated AFM tip, a Pt layer was coated on a zigzag rough surface of silicon substrate. Ultrasonic waves of frequency about 41 kHz produced a unidirectional current of about 0.15 nA and an open-circuit voltage of 0.7 mV. Although the output voltage is less as compared with the earlier work with the AFM probe tip, the deformation in ZnO nanowires is less by the ultrasonic waves.

Furthermore, to explore the new technologies working at low-frequency signals for harvesting the energy from footsteps, heartbeat, air flow, waves, noise, etc., a simple and low-cost method was introduced by Zhong Lin Wang and group to convert the low-frequency signals into electricity by using a microfiber-nanowire hybrid structure based PENG [37]. Herein, ZnO nanowires were radially grown on the surface of the microfiber textile using the hydrothermal approach [38] by entangling two fibers together, one is coated with gold and the other is not treated, and brushing them with respect to each other to produce electricity. An output voltage of 1–3 mV, output current of 4 nA and power density of 20–80 mA m^2 are generated through this method.

Other than ZnO, many materials have been studied for their piezoelectric properties. A typical example is molybdenum disulfide (MoS_2). Applying a strain of 0.53% on a single monolayer MoS_2 flake generates a peak output voltage of 15 mV, peak current of 20 pA and power density of 2 mW m^2 and an energy conversion efficiency of 5.08% [39].

An extensive number of materials have been studied for piezoelectric devices, including polyvinyldene fluoride (PVDF) [40,41], barium titanate ($BaTiO_3$) [42], cadmium sulfide (CdS) [43] and gallium nitride (GaN) [44,45]. Various flexible PENG with a wide choice of advanced materials have been developed for broad application.

Furthermore, carbon-based materials have been used due to their high conductivity and good optical, mechanical and electrical properties. Carbon-piezoelectric integrated hybrid devices were fabricated for transparent, robust device applications. Choi et al. demonstrated a transparent and flexible piezoelectric nanogenerator using a networked CNT electrode to improve device performance [46].

Moreover, Sang-Woo Kim's group investigated chemical vapor deposition (CVD)–grown large-scale graphene sheets which have been used as transparent electrodes to realise fully rollable transparent (RT) piezoelectric energy harvesting nanodevices [47].

In 2012, Kwon et al. demonstrated a PZT nanoribbon-based nanogenerator using graphene transparent electrode. A P-type doped graphene film was used to improve the PZT-graphene interface under mechanical force. The nanogenerator shows a high-output open-circuit voltage of 2 V, a current density of 2.2 μA cm^{-2} and investigated a power density of 88 mW cm^{-3} under an applied force of 0.9 kgf [48].

9.2.1.2 Geometrical Configuration

Various structural configurations of integrated nanowire array are helpful to enhance the output performance of PENG. The three types of geometrical configurations are as follows:

  I. Lateral-nanowire integrated nanogenerator (LING)
 II. Vertical-nanowire integrated nanogenerator (VING)
III. Nanocomposite electrical generator (NEG)

  I. *Lateral-nanowire integrated nanogenerator (LING)*: This includes a two-dimensional configuration [49] and consists of a base electrode, a laterally grown piezoelectric nanostructure and a metal electrode for Schottky contacts [21]. Parallel nanowires grown on the flexible substrate are the expansion of the single-wire generator (Figure 9.4) [50]. This kind of configuration helps to harvest strain energy and large-scale nanowire integration. It involves expensive steps such as sputtering of gold or platinum, lithographic lift-off process.

Hu et al. has demonstrated a LING-type configured nanogenerator (as shown in inset of Figure 9.5) with a peak power output of 11 mW cm^3 and effective energy conversion of 4.6% [21].

 II. *Vertical-nanowire integrated nanogenerator (VING)*: This has a three-dimensional geometrical configuration. It consists of an electrode, nanowire array and counter-electrode. The topology of the counter-electrode is of

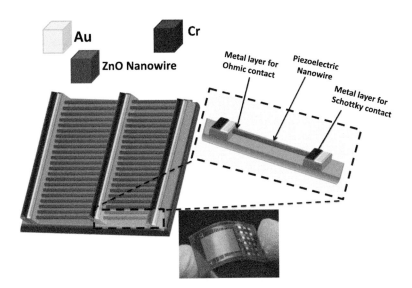

FIGURE 9.4 Lateral-nanowire integrated nanogenerator (LING) configuration. Inset shows flexible high-output nanogenerator based on lateral ZnO nanowire array. (Reproduced with permission from G. Zhu et al., *Nano Lett.* 10, 2010, 3151–3155. doi: 10.1021/nl101973h)

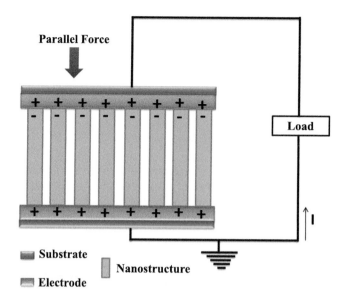

FIGURE 9.5 Vertical-nanowire integrated nanogenerator.

great importance. VING with partial contact generates a direct current (DC) signal, whereas full contact provides alternating current (Figure 9.5) [51]. A VING-configuration based nanogenerator involves low cost and easy synthesis steps. In 2007, Wang et al. introduced an initial VING-configured nanodevice wherein the deformation of a single vertical ZnO nanowire occurred via mechanical stress by the AFM tip [36]. In 2009, a vertically aligned ZnO nanorod array on ITO/PES substrate was demonstrated to generate an output current density of 1 μA cm^{-2} [30].

III. *Nanocomposite electrical generator (NEG)*: It is also three-dimensionally configured and consists of metal-plate electrode, vertically grown nanostructures and polymer matrix to fill the gap. It requires a complicated filling process and expensive solutions.

Momeni et al., in 2010, designed a nanocomposite electrical generator that is composed of a finite epoxy embedded ZnO nanowires array [52]. Further, using continuum mechanics and Maxwell's equations, they have been provided with the modelling of a NEG [53]. Linear piezoelectric behaviour has been assumed with perfect bonding between ZnO NWs and polymer matrix.

9.2.1.3 Formulae Used

Calculation of applied pressure (σ) and current density (J_{SC}) evaluation: By combining the physical models of gravity and pulse term, we can calculate the pressure imparted by a human finger [54]. When the human finger strikes the surface of the device, it initially touches the surface of the device film and then completely acts all over the surface of the film. Equations 9.1 through 9.4 are used to calculate the imparted stress according to the energy and momentum conservation principles:

Nanogenerator

$$m \cdot g \cdot h = \frac{1}{2} m \cdot v^2 \quad (9.1)$$

$$(F - m \cdot g) \cdot \Delta t = m \cdot v \quad (9.2)$$

$$\sigma = \frac{F}{S} \quad (9.3)$$

$$J_{SC} = \frac{I_{SC}}{S} \quad (9.4)$$

where m is the object mass, h is the height or distance between the human finger and the device surface, v is the maximum falling speed, σ refers to applied pressure, g is gravity (9.8 N kg^{-1}), F is the contact force, S belongs to the effective contact area, Δt is the time span when the second process take place, J_{SC} is current density (μA cm^{-2}) and I_{SC} is the short-circuit current in μA.

The released energy density (U_R) at any electric field can be calculated by integration of the area between the polarisation P versus electric field E loop and electric field co-ordinate; the total energy density (U_T) stored in the nanocomposite film is the addition of released energy density and the energy loss density of the nanocomposite [55].

The efficiency of the nanocomposite can be obtained from the ratio between released energy density (U_R) and total energy density (U_T) as written in Equation 9.5:

$$\eta = \frac{U_R}{U_T} \quad (9.5)$$

9.2.1.4 Synthesis Methodologies

Various synthesis techniques have been used for the synthesis of piezoelectric nanomaterials. They are categorised into two different approaches: bottom-up approach and top-down approach.

I. *Bottom-up approach*: Sol-gel processes [56], laser pyrolysis [57,58], CVD technique [59], spraying fabrication process [60], chemical bath deposition (CBD), hydrothermal methods [61], aqueous chemical growth (ACG)[62].
II. *Top-down approaches*: They include nanolithography technique [63], anodisation process [64], laser machining [65] and dry etching [66].

9.2.1.5 Factors Affecting Performance of Piezoelectric Nanogenerator

The important factors that help to enhance the performance of nanogenerators are as follows:

1. Carrier density [67]
2. Height of Schottky barrier [68]
3. Electrode conductivity
4. Quality, length and contact area of NWs [69]

9.2.2 PYROELECTRIC NANOGENERATOR

PYNG is an emerging energy harvester to scavenge the most abundant and widely available waste heat from our living environment and industrial waste [9]. It is able to convert thermal energy into electrical energy based on the pyroelectric effect. According to the pyroelectric effect, the thermal gradient creates spontaneous polarisation in an anisotropic solid and induces charge separation along size-driven enhancement in the pyroelectric coupling.

9.2.2.1 Mechanism

The operational mechanism of pyroelectric nanogenerators can be explained in two different ways:

I. *Primary pyroelectric effect*: It involves
   - Strain-free case
   - Domination of pyroelectric effect in PZT, $BaTiO_3$ and in some other ferroelectric materials
   - Random orientation of electric dipoles around its equilibrium axis induced by thermal fluctuations
   - Magnitude of dipoles wobbling increases with the increase in temperature

Yang et al. introduced the pyroelectric nanogenerator using lead-free $KNbO_3$ nanowires. A composite material of single crystalline $KNbO_3$ nanowires in the growth direction of [011] with polydimethylsiloxane (PDMS) polymer has been used to make a flexible nanogenerator [70]. The device consists of three stacked layers: a top silver Ag electrode, a $KNbO_3$ NW-PDMS composite film and an indium tin oxide (ITO) film as the bottom electrode. The diameter of the perovskite $KNbO_3$ NW is about 150 nm, and the growth direction is along [011]. The PYNG gives 2 mV output voltage and 20 pA current under the cyclic change of the temperature from 295–298 K.

II. *Secondary pyroelectric effect*: It involves
   - Strain-induced anisotropic thermal expansion
   - Domination of the pyroelectric response in ZnO, CdS and asymmetric wurtzite-type materials
   - Piezopotential induces due to thermal deformation
   - Output associated with piezoelectric co-efficient and anisotropic thermal expansion of the material

Yang et al. demonstrated a pyroelectric nanogenerator with excellent stability of the ZnO-nanowire, and the characteristic co-efficient of heat flow conversion into electricity is about 50–80 mV m^2 W^{-1} (Figure 9.6) [71]. Herein, the schematic shows Ag and ITO as top and bottom electrodes, respectively, and ZnO NWs array as the pyroelectric material. PYNG produces an output voltage of 5.8 mV and a current of 108.5 pA with the change in temperature 294–305 K periodically.

A lightweight and robust self-sufficient powered hybrid NG based on piezoelectric and pyroelectric effects of the electrospun nonwoven poly(vinylidene fluoride)

Nanogenerator

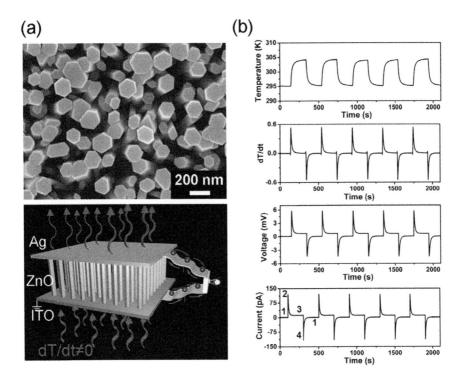

FIGURE 9.6 The mechanism of the pyroelectric nanogenerator based on the secondary pyroelectric effect. (a) Scanning electron microscope image of ZnO NW array and schematic diagram of the structure of a pyroelectric nanogenerator. (b) The output voltage and current of the device under the cyclic change in temperature. (Reproduced with permission from Y. Yang et al., *Nano Lett.* 12, 2012, 2833–2838. doi: 10.1021/nl3003039)

(PVDF) nanofiber membrane (NFM) acts as an active layer without any post-poling treatment, was introduced by Ming-Hao You and group in 2018 [72].

9.2.2.2 Formulae Used

Usually, the output current I of the pyroelectric NGs can be determined by Equation 9.6:

$$I = p \cdot A \cdot \frac{dT}{dt} \quad (9.6)$$

where I is pyroelectric current, P is pyroelectric co-efficient, A is the effective surface area and dT/dT is the rate of change of temperature with time [73].

Pyroelectric nanogenerators can be applied in various applications where a time-dependent temperature fluctuation exists. A self-sufficient powered temperature sensor without a battery is known to be a feasible application of the pyroelectric nanogenerator where the response time and reset time of the sensor are about 0.9 and 3 s, respectively [74].

9.2.3 TRIBOELECTRIC NANOGENERATOR

TENG harvests mechanical motion into electrical signals on the basis of two conjugated mechanisms: triboelectrification and electrostatic induction effects.

9.2.3.1 Mechanism

The triboelectric effect is a typical phenomenon in which a material is electrically charged as it contacts another triboactive material through friction. More progressive studies are still going on to understand the mechanism of triboelectrification. When two different materials come into contact with each other, adhesion between the two surfaces of the materials gives rise to chemical bonding. The bonding forms a pathway for the transportation of charges from one material to another and balances the electrochemical potential. When the surfaces separate, a few of the bonded atoms have a tendency to acquire the extra electrons and a few of them give them away and produce triboelectric charges at the surfaces. The electrons are forced to flow through the external circuit by the present triboelectric charges on the dielectric in order to neutralise the electric potential difference between the electrodes. Based on the effective principle, four different modes of TENGs are elaborated as follows:

I. *Vertical-contact separation mode*: Involves two dissimilar vertically aligned dielectric films, top and bottom contact electrodes. Physical contact between the dielectric films generates the electric charges and a contact potential difference between the dielectric films. It builds an opposite potential to balance the electrostatic field (Figure 9.7a) [75].

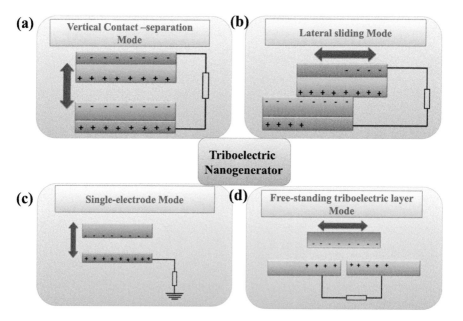

FIGURE 9.7 (a) Vertical contact separation mode, (b) lateral-sliding mode, (c) single-electrode mode and (d) free-standing triboelectric layer mode.

II. *Lateral sliding mode*: Involves two dissimilar dielectric films relatively sliding in parallel to the surface and generates lateral polarisation along the sliding direction (Figure 9.7b) [76]. It gives AC output signals. The sliding can be planar motion, disc rotation [77] or cylindrical rotation [78].
III. *Single-electrode mode*: Includes one triboelectric electrode material, and the bottom part of the electrode is grounded (Figure 9.7c). Exchange of electrons was carried out between the bottom electrode and the ground to balance the potential difference. This mode can harvest energy in both contact separation mode [79] and contact sliding mode [80]. A human's skin can also be considered as a grounded electrode.
IV. *Free-standing triboelectric layer mode*: Involves harvesting energy from moving objects such as a human walking, vehicles, running trains, etc., consisting of a pair of symmetric metal electrodes underneath a dielectric layer (Figure 9.7d), high energy conversion efficiency and robustness [81].

9.2.3.2 Formulae Used

The open-circuit potential (V_{OC}) can be calculated using Equation 9.7:

$$V_{OC} = -\frac{\sigma \cdot d}{\varepsilon_0} \tag{9.7}$$

where σ is the triboelectric charge density, ε_0 is the vacuum permittivity and d is the gap between the two contact materials.

9.2.3.3 Boosting Triboelectric Nanogenerator Performance

Output performances such as output power, current and energy conversion efficiency of the TENG must be improved for commercialisation. Hence, suitable principles and mechanisms have been explored to realise the enhancement of TENGs as a sustainable power source for electronics. For existing triboelectric nanogenerators (TENGs), it is important to find unique methods to further enhance the output power.

- *Introducing electric double layer*: Chun et al. performed a practical experiment to boost the output performance of TENG via introducing an electric double-layer effect. They fabricated a portable power supply system composed of three layers: an electric double layer is inserted between a top layer, made of Al/polydimethylsiloxane (PDMS), and a bottom layer, made of Al. As the force is applied from the top and released, the opposite charge distribution occurs in the middle layer by a sequential contact configuration and straight electrical connection of the center layer to the earth, forms an electric double layer and enhances the electric potential value. A sustainable and 16-fold enhanced output current and power performance of 1.22 mA and 46.8 mW cm^2 under a low frequency of 3 Hz and a force of 50 N is produced with an energy conversion efficiency of 22.4% [82].
- *Surface texturing and dielectric properties*: By modifying the surface texture and controlling the dielectric constant, the output performance of PDMS-based TENGs can be boosted which becomes a promising strategy

for high-performance triboelectric nanogenerators. Introducing highly dielectric particles significantly helps in enhancement of the dielectric constant and surface charge potential of the PDMS film, supported by both experimental analysis and COMSOL simulation. A 10-fold increase in power output as compared with flat PDMS-based TENGs with an output voltage of ~390 V (peak to peak), a short-circuit current density of ~170 mA m^{-2}, and a charge density of ~108 μ cm^{-2} were obtained, corresponding to the peak power density of 9.6 W m^{-2} [83].

- *Restoration of charges*: Guo et al. introduced a new method to improve the performance of TENG via charge replenishment by introducing a frictional rod in the structure design. With grating number 30 and a gap of 0.5 mm, fluorinated ethylene propylene (FEP)–based TENG delivers an output power of 250 mW m^{-2} at a rotating rate of rolling friction rod 1000 rotation/minute [84].

9.2.3.4 Recent Progress in Advancement in Triboelectric Nanogenerator Applications

Nanogenerators are promising for the miniaturisation of power packages and the self-powering of nanosystems used in implantable bio-sensing, environmental monitoring and personal electronics. Recently, research has been focussed on the integration of nanogenerators with solar cells and supercapacitors or directly with sensors. These hybrid systems provide sustainable power for electronic applications.

The major applications of TENG can be classified into four different areas:

I. Sustainable micropower sources for self-powered systems
II. Active sensors for medical; tactile touch sensors, acoustic sensors, acceleration sensors, chemical sensors, etc.
III. Infrastructure, environmental monitoring
IV. Human-machine interfacing (HMI)

Various harvesting energy resources have been used to explore numerous applications of TENGs:

- Biomechanical energy such as human walking, heartbeats, machine vibrations, wind, knee and arm bending, etc.
- Blue energy such as wind energy, ultrasonic waves, water waves, raindrop energy, etc.
- Direct high-voltage power sources

9.4 CARBON-BASED NANOGENERATORS

As explained earlier, the advantage of using carbonaceous material as an electrode material as well as a current collector by replacing metals in nanogenerators, makes it a promising candidate in the era of mechanical energy harvesting. A number of reports have been published on carbon-based nanogenerators. A comparison of the performance of various carbon-based nanogenerators is given in Table 9.1.

TABLE 9.1
Comparison of Carbon-Based Nanogenerators' Performances Reported in Literature

Sr. No.	Electrode	Method	Open-Circuit Voltage (V_{OC})	Current (I_{SC})	Power Density (P_{max})	Ref.
1.	3D-Carbon electrode/PTFE	Ultraviolet-lithography	75 V	35 μA	0.35 mW cm^{-2}	[85]
2.	rGO	Electrochemical process	7 mV	4 μA	–	[86]
3.	Graphene	Chemical vapor deposition (CVD) method	5 V	500 nA cm^{-2}	–	[87]
4.	Graphene/PTFE/PDMS	Thermal CVD method	40 V	3.9 μA	130 μW	[88]
5.	Crumpled graphene	Commercially purchased	83 V	25.78 μA	0.25 mW cm^{-2}	[89]
6.	Graphene/ZnO nanorods	CVD grown graphene	–	2 μA cm^{-2}	–	[47]
7.	PZT-ribbons/graphene	Sol–gel method	2 V	2.2 μA cm^{-2}	88 mW cm^{-3}	[48]
8.	Carbon electrode materials	Chemical method	1300 V	1.2 mA	74 W cm^{-2}	[90]
9.	Polyimide/rGO	Thermal annealing	130 V	7.5 μA	–	[91]
10.	Graphene	CVD grown	650 V	12 μA	3.28 mW	[13]
11.	rGO/PVDF/TrFE	Scrap coating	4 V	0.6 μA	28 W m^{-3}	[92]

9.5 CONCLUSIONS

Research in the field of TENG moves a step forward in converting mechanical energy into electricity to fulfil human living demands. High-output voltage and power density, large conversion efficiency, light weight, lower price, ease of fabrication, simple maintenance process and small-scale size make nanogenerators an exciting technology. Moreover, the extension of TENG with other emerging technologies such as the solar cell to simultaneously harvest various other forms of energies proves their potential for many applications in the IOTs, mobile electronics, self-powered e-skin technology, fabric electronics, environmental protection and medical monitoring machines.

In addition, the pocket-size TENG can be packaged with many small size units having high-output current and voltage without reducing the power. Much effort has been spent on improving the durability and output stability of the device. The TENG with energy storage systems have become a new paradigm in energy technologies for truly achieving sustainable and maintenance-free, self-powered systems.

REFERENCES

1. C. Liu, F. Li, L.-P. Ma, H.-M. Cheng, Advanced materials for energy storage, *Adv. Mater.* 22, 2010, E28–E62. doi: 10.1002/adma.200903328
2. Z.L. Wang, J. Song, Piezoelectric nanogenerators based on zinc oxide nanowire arrays, *Science* 312, 2006, 242–246. doi: 10.1126/science.1124005
3. S. Beeby, M.J. Tudor, N. White, Energy harvesting vibration sources for microsystems applications, *Meas. Sci. Technol.* 17, 2006. doi: 10.1088/0957-0233/17/12/R01
4. D. Li, G. Hu, Nanoscale Biosensors BT – Encyclopedia of Microfluidics and Nanofluidics, in: D. Li, editor, Springer US, Boston, MA, 2008: pp. 1447–1448. doi: 10.1007/978-0-387-48998-8_1103
5. A. Cavalcanti, B. Shirinzadeh, R.A. Freitas Jr, T. Hogg, Nanorobot architecture for medical target identification, *Nanotechnology* 19, 2008, 015103. doi: 10.1088/0957-4484/19/01/015103
6. Z. Liu, Y. Liu, Y. Chang, H.R. Seyf, A. Henry, A.L. Mattheyses, K. Yehl, Y. Zhang, Z. Huang, K. Salaita, Nanoscale optomechanical actuators for controlling mechanotransduction in living cells, *Nat. Methods.* 13, 2015, 143.
7. P. Glynne-Jones, N. White, Self-powered systems: A review of energy sources, *Sens. Rev.* 21, 2001. doi: 10.1108/02602280110388252
8. Z. Li, J. Shen, I. Abdalla, J. Yu, B. Ding, Nanofibrous membrane constructed wearable triboelectric nanogenerator for high performance biomechanical energy harvesting, *Nano Energy* 36, 2017, 341–348. doi: 10.1016/j.nanoen.2017.04.035
9. H. Xue, Q. Yang, D. Wang, W. Luo, W. Wang, M. Lin, D. Liang, Q. Luo, A wearable pyroelectric nanogenerator and self-powered breathing sensor, *Nano Energy*, 38, 2017. doi: 10.1016/j.nanoen.2017.05.056
10. Y. Hu, C. Xu, Y. Zhang, L. Lin, R. L Snyder, Z. Wang, A nanogenerator for energy harvesting from a rotating tire and its application as a self-powered pressure/speed sensor, *Adv. Mater.* 23, 2011. doi: 10.1002/adma.201102067
11. Z.L. Wang, Triboelectric nanogenerators as new energy technology and self-powered sensors – Principles, problems and perspectives, *Faraday Discuss.* 176, 2014, 447–458. doi: 10.1039/C4FD00159A
12. Y. Wang, Y. Yang, Z.L. Wang, Triboelectric nanogenerators as flexible power sources, *Npj Flex. Electron.* 1, 2017, 10. doi: 10.1038/s41528-017-0007-8
13. S. Ankanahalli Shankaregowda, B.N. Chandrashekar, X. Cheng, M. Shi, Z. Liu, H.-X. Zhang, A flexible and transparent graphene based triboelectric nanogenerator, *IEEE Trans. Nanotechnol.* 15, 2016, 435–441. doi: 10.1109/TNANO.2016.2540958.
14. Z. Li, J. Shen, I. Abdalla, J. Yu, B. Ding, Nanofibrous membrane constructed wearable triboelectric nanogenerator for high performance biomechanical energy harvesting, *ARPN J. Eng. Appl. Sci.* 8, 2013, 504–512.
15. L.C. Rome, L. Flynn, E.M. Goldman, T.D. Yoo, Generating electricity while walking with loads, *Science* 309, 2005, 1725–1728. doi: 10.1126/science.1111063
16. X. Bai, Y. Wen, J. Yang, P. Li, J. Qiu, Y. Zhu, A magnetoelectric energy harvester with the magnetic coupling to enhance the output performance, *J. Appl. Phys.* 111, 2012, 07A938. doi: 10.1063/1.3677877
17. Z. Lin, J. Chen, X. Li, J. Li, J. Liu, Q. Awais, J. Yang, Broadband and three-dimensional vibration energy harvesting by a non-linear magnetoelectric generator, *Appl. Phys. Lett.* 109, 2016, 253903. doi: 10.1063/1.4972188
18. P.D. Mitcheson, P. Miao, B.H. Stark, E.M. Yeatman, A.S. Holmes, T.C. Green, MEMS electrostatic micropower generator for low frequency operation, *Sensors Actuators A Phys.* 115, 2004, 523–529. doi: https://doi.org/10.1016/j.sna.2004.04.026
19. H.R. Foruzande, A. Hajnayeb, A. Yaghootian, Nanoscale piezoelectric vibration energy harvester design, *AIP Adv.* 7, 2017, 95122. doi: 10.1063/1.4994577

20. S. Xu, B.J. Hansen, Z.L. Wang, Piezoelectric-nanowire-enabled power source for driving wireless microelectronics., *Nat. Commun.* 1, 2010, 93. doi: 10.1038/ncomms1098
21. G. Zhu, R. Yang, S. Wang, Z.L. Wang, Flexible high-output nanogenerator based on lateral ZnO nanowire array, *Nano Lett.* 10, 2010, 3151–3155. doi: 10.1021/nl101973h
22. B. Kumar, K.Y. Lee, H.-K. Park, S.J. Chae, Y.H. Lee, S.-W. Kim, Controlled growth of semiconducting nanowire, nanowall, and hybrid nanostructures on graphene for piezoelectric nanogenerators, *ACS Nano.* 5, 2011, 4197–4204. doi: 10.1021/nn200942s
23. R.A. Whiter, V. Narayan, S. Kar-Narayan, A scalable nanogenerator based on self-poled piezoelectric polymer nanowires with high energy conversion efficiency, *Adv. Energy Mater.* 4, 2014, 1400519. doi: 10.1002/aenm.201400519
24. S. Crossley, S. Kar-Narayan, Energy harvesting performance of piezoelectric ceramic and polymer nanowires, *Nanotechnology* 26, 2015, 344001. doi: 10.1088/0957-4484/26/34/344001
25. F.L. Boughey, T. Davies, A. Datta, R.A. Whiter, S.-L. Sahonta, S. Kar-Narayan, Vertically aligned zinc oxide nanowires electrodeposited within porous polycarbonate templates for vibrational energy harvesting., *Nanotechnology* 27, 2016, 28LT02. doi: 10.1088/0957-4484/27/28/28LT02
26. Q. Zhang, E.S. Kim, Vibration energy harvesting based on magnet and coil arrays for watt-level handheld power source, *Proc. IEEE.* 102, 2014, 1747–1761. doi: 10.1109/JPROC.2014.2358995
27. R.J. de Graaff, K.T. Compton, L.C. Van Atta, The electrostatic production of high voltage for nuclear investigations, *Phys. Rev.* 43, 1933, 149–157. doi: 10.1103/PhysRev.43.149
28. D. Shen, J.-H. Park, J.H. Noh, S.-Y. Choe, S.-H. Kim, H.C. Wikle, D.-J. Kim, Micromachined PZT cantilever based on SOI structure for low frequency vibration energy harvesting, *Sensors Actuators A Phys.* 154, 2009, 103–108. doi: https://doi.org/10.1016/j.sna.2009.06.007
29. S. Katzir, The discovery of the piezoelectric effect, *Arch. Hist. Exact Sci.* 57, 2003, 61–91. doi: 10.1007/s00407-002-0059-5
30. M.-Y. Choi, D. Choi, M.-J. Jin, I. Kim, S.-H. Kim, J.-Y. Choi, S.Y. Lee, J.M. Kim, S.-W. Kim, Mechanically powered transparent flexible charge-generating nanodevices with piezoelectric ZnO nanorods, *Adv. Mater.* 21, 2009, 2185–2189. doi: 10.1002/adma.200803605
31. Y. Suzuki, Recent progress in MEMS electret generator for energy harvesting, *IEEJ Trans. Electr. Electron. Eng.* 6, 2011, 101–111. doi: 10.1002/tee.20631
32. T. Sterken, P. Fiorini, K. Baert, R. Puers, G. Borghs, An electret-based electrostatic μ-generator, *Transducers '03. 12th International Conference on Solid-State Sensors, Actuators and Microsystems*, Boston, MA, June 8–12, 2003. doi: 10.1109/SENSOR.2003.1217009
33. S. Matsusaka, H. Maruyama, T. Matsuyama, M. Ghadiri, Triboelectric charging of powders: A review, *Chem. Eng. Sci.* 65, 2010, 5781–5807. doi: https://doi.org/10.1016/j.ces.2010.07.005
34. X.-S. Zhang, M.-D. Han, B. Meng, H.-X. Zhang, Review, *Nano Energy.* 11, 2015, 304–322. doi: 10.1016/j.nanoen.2014.11.012
35. R. Wen, J. Guo, A. Yu, K. Zhang, J. Kou, Y. Zhu, Y. Zhang, B.-W. Li, J. Zhai, Remarkably enhanced triboelectric nanogenerator based on flexible and transparent monolayer titania nanocomposite, *Nano Energy.* 50, 2018, 140–147. doi: 10.1016/j.nanoen.2018.05.037
36. X. Wang, J. Song, J. Liu, Z.L. Wang, Direct-current nanogenerator driven by ultrasonic waves, *Science* 316, 2007, 102–105. doi: 10.1126/science.1139366
37. Y. Qin, X. Wang, Z.L. Wang, Microfibre-nanowire hybrid structure for energy scavenging, *Nature* 451, 2008, 809–813. doi: 10.1038/nature06601
38. J.W.P. Hsu, Z.R. Tian, N.C. Simmons, C.M. Matzke, J.A. Voigt, J. Liu, Directed spatial organization of zinc oxide nanorods, *Nano Lett.* 5, 2005, 83–86. doi: 10.1021/nl048322e

39. W. Wu, L. Wang, Y. Li, F. Zhang, L. Lin, S. Niu, D. Chenet et al., Piezoelectricity of single-atomic-layer MoS_2 for energy conversion and piezotronics, *Nature* 514, 2014, 470–474. doi: 10.1038/nature13792
40. C. Chang, V.H. Tran, J. Wang, Y.-K. Fuh, L. Lin, Direct-write piezoelectric polymeric nanogenerator with high energy conversion efficiency, *Nano Lett.* 10, 2010, 726–731. doi: 10.1021/nl9040719
41. Y. Zi, L. Lin, J. Wang, S. Wang, J. Chen, X. Fan, P.-K. Yang, F. Yi, Z.L. Wang, Triboelectric–pyroelectric–piezoelectric hybrid cell for high-efficiency energy-harvesting and self-powered sensing, *Adv. Mater.* 27, 2015, 2340–2347. doi: 10.1002/adma.201500121
42. Z. Wang, J. Hu, A.P. Suryavanshi, K. Yum, M.-F. Yu, Voltage generation from individual $BaTiO_3$ nanowires under periodic tensile mechanical load, *Nano Lett.* 7, 2007, 2966–2969. doi: 10.1021/nl070814e
43. Y.-F. Lin, J. Song, Y. Ding, S.-Y. Lu, Z.L. Wang, Piezoelectric nanogenerator using CdS nanowires, *Appl. Phys. Lett.* 92, 2008, 22105. doi: 10.1063/1.2831901
44. C.-Y. Chen, G. Zhu, Y. Hu, J.-W. Yu, J. Song, K.-Y. Cheng, L.-H. Peng, L.-J. Chou, Z.L. Wang, Gallium nitride nanowire based nanogenerators and light-emitting diodes, *ACS Nano.* 6, 2012, 5687–5692. doi: 10.1021/nn301814w
45. C.-T. Huang, J. Song, W.-F. Lee, Y. Ding, Z. Gao, Y. Hao, L.-J. Chen, Z.L. Wang, GaN nanowire arrays for high-output nanogenerators, *J. Am. Chem. Soc.* 132, 2010, 4766–4771. doi: 10.1021/ja909863a
46. D. Choi, M.-Y. Choi, H.-J. Shin, S.-M. Yoon, J.-S. Seo, J.-Y. Choi, S.Y. Lee, J.M. Kim, S.-W. Kim, Nanoscale networked single-walled carbon-nanotube electrodes for transparent flexible nanogenerators, *J. Phys. Chem. C.* 114, 2010, 1379–1384. doi: 10.1021/jp909713c
47. D. Choi, M.-Y. Choi, W.M. Choi, H.-J. Shin, H.-K. Park, J.-S. Seo, J. Park et al., Fully rollable transparent nanogenerators based on graphene electrodes, *Adv. Mater.* 22, 2010, 2187–2192. doi: 10.1002/adma.200903815
48. J. Kwon, W. Seung, B.K. Sharma, S.-W. Kim, J.-H. Ahn, A high performance PZT ribbon-based nanogenerator using graphene transparent electrodes, *Energy Environ. Sci.* 5, 2012, 8970–8975. doi: 10.1039/C2EE22251E
49. R. Yang, Y. Qin, L. Dai, Z.L. Wang, Power generation with laterally packaged piezoelectric fine wires, *Nat. Nanotechnol.* 4, 2009, 34–39. doi: 10.1038/nnano.2008.314
50. S. Xu, Y. Qin, C. Xu, Y. Wei, R. Yang, Z.L. Wang, Self-powered nanowire devices, *Nat. Nanotechnol.* 5, 2010, 366
51. M. Ani Melfa Roji, G. Jiji, T. Ajith Bosco Raj, A retrospect on the role of piezoelectric nanogenerators in the development of the green world, *RSC Adv.* 7, 2017, 33642–33670. doi: 10.1039/C7RA05256A
52. K. Momeni, G.M. Odegard, R.S. Yassar, Nanocomposite electrical generator based on piezoelectric zinc oxide nanowires, *J. Appl. Phys.* 108, 2010, 114303. doi: 10.1063/1.3517095
53. K. Momeni, S.M.Z. Mortazavi, Optimal aspect ratio of zinc oxide nanowires for a nanocomposite electrical generator, *J. Comput. Theor. Nanosci.* 9, 2012, 1670–1674. doi: 10.1166/jctn.2012.2262
54. F. Huang, Z. Wang, X. Lu, J. Zhang, K. Min, W. Lin, R. Ti, T. Xu, J. He, C. Yue, J. Zhu, Peculiar magnetism of $BiFeO_3$ nanoparticles with size approaching the period of the spiral spin structure, *Sci. Rep.* 3, 2013, 2907. doi: 10.1038/srep02907
55. S.K. Karan, D. Mandal, B.B. Khatua, Self-powered flexible Fe-doped RGO/PVDF nanocomposite: An excellent material for a piezoelectric energy harvester, *Nanoscale* 7, 2015, 10655–10666. doi: 10.1039/C5NR02067K
56. H. Bahadur, A.K. Srivastava, D. Haranath, H. Chander, A. Basu, S.B. Samanta, K.N. Sood, R. Kishore, Nano-structured ZnO films by sol-gel process, *Int. J. Pure Appl. Phys.* 45, 2007, 395–399

57. M.G.S. Bernd, S.R. Bragança, N. Heck, L.C.P. da Silva Filho, Synthesis of carbon nanostructures by the pyrolysis of wood sawdust in a tubular reactor, *J. Mater. Res. Technol.* 6, 2017, 171–177. doi: https://doi.org/10.1016/j.jmrt.2016.11.003
58. T. J. Mckinnon, A. M. Herring, B. D. Mccloskey, Laser pyrolysis method for producing carbon nano-spheres, 2009. Patent No. WO2005023708A2.
59. S. Manna, J.W. Kim, Y. Takahashi, O.G. Shpyrko, E.E. Fullerton, Synthesis of single-crystalline anisotropic gold nano-crystals via chemical vapor deposition, *J. Appl. Phys.* 119, 2016, 174301. doi: 10.1063/1.4948565
60. L.J. and F. Gitzhofer, Induction plasma synthesis of nanostructured SOFCs electrolyte using solution and suspension plasma spraying: A comparative study, *J. Therm. Spray Technol.* 19, 2010, 566–574.
61. J. Chung, J. Lee, S. Lim, Annealing effects of ZnO nanorods on dye-sensitized solar cell efficiency, *Phys. B Condens. Matter.* 405, 2010, 2593–2598. doi: https://doi.org/10.1016/j.physb.2010.03.041
62. M.M. Yusoff, M.H. Mamat, M.F. Malek, A.B. Suriani, A. Mohamed, M.K. Ahmad, S.A.H. Alrokayan, H.A. Khan, M. Rusop, Growth of titanium dioxide nanorod arrays through the aqueous chemical route under a novel and facile low-cost method, *Mater. Lett.* 164, 2016, 294–298. doi: https://doi.org/10.1016/j.matlet.2015.11.014
63. E. Di Fabrizio, R. Fillipo, S. Cabrini, R. Kumar, F. Perennes, M. Altissimo, L. Businaro, et al., X-ray lithography for micro- and nano-fabrication at ELETTRA for interdisciplinary applications, *J. Phys. Condens. Matter.* 16, 2004, S3517–S3535. doi: 10.1088/0953-8984/16/33/013
64. Y. Yang, Y. Xia, W. Huang, J. Zheng, Z. Li, Fabrication of nano-network gold films via anodization of gold electrode and their application in SERS, *J. Solid State Electr.* 16, 2011. doi: 10.1007/s10008-011-1600-8
65. K. Kim, Y.-M. Choi, D.-G. Gweon, M.G. Lee, A novel laser micro/nano-machining system for FPD process, *J. Mater. Process. Technol.* 201, 2008. doi: 10.1016/j.jmatprotec.2007.11.186
66. I. Rangelow, Dry etching-based silicon micro-machining for MEMS, *Vacuum* 62, 2001. doi: 10.1016/S0042-207X(00)00442-5
67. J. Liu, P. Fei, J. Zhou, R. Tummala, Z.L. Wang, Toward high output-power nanogenerator, *Appl. Phys. Lett.* 92, 2008, 173105. doi: 10.1063/1.2918840
68. J. Liu, P. Fei, J. Song, X. Wang, C. Lao, R. Tummala, Z.L. Wang, Carrier density and Schottky barrier on the performance of DC nanogenerator., *Nano Lett.* 8, 2008, 328–332. doi: 10.1021/nl0728470
69. M. Riaz, J. Song, O. Nur, Z.L. Wang, M. Willander, Study of the piezoelectric power generation of ZnO nanowire arrays grown by different methods, *Adv. Funct. Mater.* 21, 2011, 628–633. doi: 10.1002/adfm.201001203
70. Y. Yang, J.H. Jung, B.K. Yun, F. Zhang, K.C. Pradel, W. Guo, Z.L. Wang, Flexible pyroelectric nanogenerators using a composite structure of lead-free KNbO(3) nanowires., *Adv. Mater.* 24, 2012, 5357–5362. doi: 10.1002/adma.201201414
71. Y. Yang, W. Guo, K.C. Pradel, G. Zhu, Y. Zhou, Y. Zhang, Y. Hu, L. Lin, Z.L. Wang, Pyroelectric nanogenerators for harvesting thermoelectric energy, *Nano Lett.* 12, 2012, 2833–2838. doi: 10.1021/nl3003039
72. M.-H. You, X.-X. Wang, X. Yan, J. Zhang, W.-Z. Song, M. Yu, Z.-Y. Fan, S. Ramakrishna, Y.-Z. Long, A self-powered flexible hybrid piezoelectric–pyroelectric nanogenerator based on non-woven nanofiber membranes, *J. Mater. Chem. A.* 6, 2018, 3500–3509. doi: 10.1039/C7TA10175A
73. S.B. Lang, S.A.M. Tofail, A.A. Gandhi, M. Gregor, C. Wolf-Brandstetter, J. Kost, S. Bauer, M. Krause, Pyroelectric, piezoelectric, and photoeffects in hydroxyapatite thin films on silicon, *Appl. Phys. Lett.* 98, 2011, 123703. doi: 10.1063/1.3571294

74. Y. Yang, Y. Zhou, J.M. Wu, Z.L. Wang, Single micro/nanowire pyroelectric nanogenerators as self-powered temperature sensors, *ACS Nano*. 6, 2012, 8456–8461. doi: 10.1021/nn303414u
75. S. Niu, S. Wang, L. Lin, Y. Liu, Y.S. Zhou, Y. Hu, Z.L. Wang, Theoretical study of contact-mode triboelectric nanogenerators as an effective power source, *Energy Environ. Sci*. 6, 2013, 3576–3583. doi: 10.1039/C3EE42571A
76. S. Wang, L. Lin, Y. Xie, Q. Jing, S. Niu, Z.L. Wang, Sliding-triboelectric nanogenerators based on in-plane charge-separation mechanism, *Nano Lett*. 13, 2013, 2226–2233. doi: 10.1021/nl400738p
77. L. Lin, S. Wang, Y. Xie, Q. Jing, S. Niu, Y. Hu, Z.L. Wang, Segmentally structured disk triboelectric nanogenerator for harvesting rotational mechanical energy, *Nano Lett*. 13, 2013, 2916–2923. doi: 10.1021/nl4013002
78. Y. Hu, J. Yang, Q. Jing, S. Niu, W. Wu, Z.L. Wang, Triboelectric nanogenerator built on suspended 3D spiral structure as vibration and positioning sensor and wave energy harvester, *ACS Nano*. 7, 2013, 10424–10432. doi: 10.1021/nn405209u
79. Y. Yang, Y.S. Zhou, H. Zhang, Y. Liu, S. Lee, Z.L. Wang, A single-electrode based triboelectric nanogenerator as self-powered tracking system, *Adv. Mater*. 25, 2013, 6594–6601. doi: 10.1002/adma.201302453
80. Y. Yang, H. Zhang, J. Chen, Q. Jing, Y.S. Zhou, X. Wen, Z.L. Wang, Single-electrode-based sliding triboelectric nanogenerator for self-powered displacement vector sensor system, *ACS Nano*. 7, 2013, 7342–7351. doi: 10.1021/nn403021m
81. S. Wang, Y. Xie, S. Niu, L. Lin, Z.L. Wang, Freestanding triboelectric-layer-based nanogenerators for harvesting energy from a moving object or human motion in contact and non-contact modes, *Adv. Mater*. 26, 2014, 2818–2824. doi: 10.1002/adma.201305303
82. J. Chun, B.U. Ye, J.W. Lee, D. Choi, C.-Y. Kang, S.-W. Kim, Z.L. Wang, J.M. Baik, Boosted output performance of triboelectric nanogenerator via electric double layer effect, *Nat. Commun*. 7, 2016, 12985.
83. Z. Fang, K.H. Chan, X. Lu, C.F. Tan, G.W. Ho, Surface texturing and dielectric property tuning toward boosting of triboelectric nanogenerator performance, *J. Mater. Chem. A*. 6, 2018, 52–57. doi: 10.1039/C7TA07696G
84. H. Guo, J. Chen, M.-H. Yeh, X. Fan, Z. Wen, Z. Li, C. Hu, Z.L. Wang, An ultrarobust high-performance triboelectric nanogenerator based on charge replenishment, *ACS Nano*. 9, 2015, 5577–5584. doi: 10.1021/acsnano.5b01830
85. D. Kim, B. Pramanick, A. Salazar, I.-W. Tcho, M.J. Madou, E.S. Jung, Y.-K. Choi, H. Hwang, 3D carbon electrode based triboelectric nanogenerator, *Adv. Mater. Technol*. 1, 2016, 1600160. doi: 10.1002/admt.201600160
86. W. Li, Y. Zhang, L. Liu, D. Li, L. Liao, C. Pan, A high energy output nanogenerator based on reduced graphene oxide, *Nanoscale* 7, 2015, 18147–18151. doi: 10.1039/C5NR04971G
87. S. Kim, M.K. Gupta, K.Y. Lee, A. Sohn, T.Y. Kim, K.-S. Shin, D. Kim, S.K. Kim, K.H. Lee, H.-J. Shin, D.-W. Kim, S.-W. Kim, Transparent flexible graphene triboelectric nanogenerators, *Adv. Mater*. 26, 2014, 3918–3925. doi: 10.1002/adma.201400172
88. H. Chu, H. Jang, Y. Lee, Y. Chae, J.-H. Ahn, Conformal, graphene-based triboelectric nanogenerator for self-powered wearable electronics, *Nano Energy* 27, 2016. doi: 10.1016/j.nanoen.2016.07.009
89. H. Chen, Y. Xu, J. Zhang, W. Wu, G. Song, Enhanced stretchable graphene-based triboelectric nanogenerator via control of surface nanostructure, *Nano Energy* 58, 2019, 304–311. doi: https://doi.org/10.1016/j.nanoen.2019.01.029
90. S. Gao, Y. Chen, J. Su, M. Wang, X. Wei, T. Jiang, Z.L. Wang, Triboelectric nanogenerator powered electrochemical degradation of organic pollutant using Pt-free carbon materials, *ACS Nano*. 11, 2017, 3965–3972. doi: 10.1021/acsnano.7b00422

91. X. Zhao, B. Chen, G. Wei, J.M. Wu, W. Han, Y. Yang, Polyimide/graphene nanocomposite foam-based wind-driven triboelectric nanogenerator for self-powered pressure sensor, *Adv. Mater. Technol.* 4, 2019, 1800723. doi: 10.1002/admt.201800723
92. X. Hu, Z. Ding, L. Fei, Y. Xiang, Y. Lin, Wearable piezoelectric nanogenerators based on reduced graphene oxide and in situ polarization-enhanced PVDF-TrFE films, *J. Mater. Sci.* 54, 2019, 6401–6409. doi: 10.1007/s10853-019-03339-5

10 Carbonaceous Materials–Based Hybrid Energy Technologies

10.1 INTRODUCTION TO HYBRID TECHNOLOGY

A hybrid energy system includes the harvesting of energy and energy storage as well as hybrid energy storage systems and technologies. Harvesting energy, storage and utilization are becoming global problems owing to very fast development in smart electronics for daily life. In recent years, the incorporation of harvesting energy and storage technology was thought to be one of the most important energy-related technologies owing to the prospect of switching batteries or at least extending the lifetime of a battery [1].

Currently, smart electronics with multiple functionalities such as mobile phones, tablets, electronic gadgets and various sensors have become pervasive everywhere in our modern lives. If it is considered that such devices are designed for high efficiency and can perform well on a battery for a long time period, often the power consumption is heavily loaded due to the relatively large size of the devices. Accordingly, sustainable power sources are extensively required for their low maintenance cost, self-governing and uninterrupted operation of such low-energy-consumption future electronics.

An electrical energy storage (EES) system plays a vibrant role in our modern life due to our dependence on various electronic devices that require flexibility. The restrictions of current hybrid EES systems have fundamental gaps in our knowledge of atomic- and molecular-level processes that manage their operation, performance and failure. In routine batteries, starting from 'air' to 'zinc', there are still demands for fundamental developments. Although 'lithium' is at the centre of new material improvements, even it requires improvement in the nature of the cathode, anode and electrolytes, as well as their structural shaping and electrode-electrolyte integration [2]. Hybridisation of the different storage devices and even hybrid materials offer openings for synergic behaviour and upgraded properties with respect to their individual components. Generally, lithium-ion batteries (LIBs) engaged for use in hybrid electric vehicles (HEVs) are intended for high power consumption based on $LiFePO_4$, in the small LIBs, but has enormous capabilities for high rate and long life cycle in the EV. Supercapacitors are electrochemical storage devices that have high-rate capability for fast charging/discharging. However, because of their low energy density, they cannot be used in HEVs. Therefore, it is an obvious demand for a hybrid energy storage system to have high energy as well as high power, which can be fulfilled by hybridisation – a battery-supercapacitor [3].

Energy storage systems including hybrid technology are very important to harvesting the energy in terms of hybrid energy technology today. Energy harvesting can be achieved through various processes including pyroelectric, piezoelectric, triboelectric, flexoelectric, thermoelectric and photovoltaic effects, which have been extensively studied for practical applications (Figure 10.1) [1,4]. The common problem with all the energy harvesting systems is that they utilise only one type of energy, with the other type wasted. Therefore, it is highly desired to develop an integrated energy harvester to accumulate various energy to convert into electricity. This hybridisation process can help to fully utilise waste energy and can power smart devices easily.

In this chapter, we describe the recent research progress on hybrid energy storage systems as well as hybrid energy harvest-storage systems and the technology associated with them. Research on various integrated muti-type energy devices as well as self-powered devices is covered to provide an understanding of the engineering and technology necessary for the integration of the systems. Carbon nanomaterials (graphene, carbon nanotubes [CNTs], activated carbon, carbon nanofiber, etc.) and their composites have several advantages including good electrical conductivity, huge surface area, high mechanical strength and lower cost making them suitable candidates for hybrid energy systems.

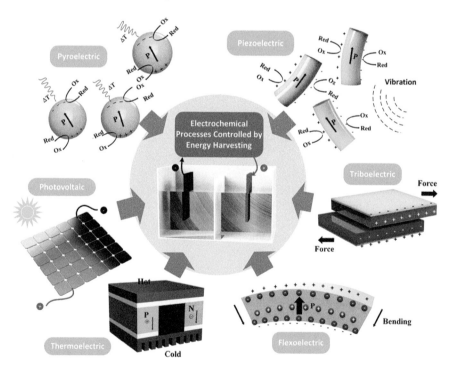

FIGURE 10.1 Pyroelectric (temperature cycles), thermoelectric (temperature gradients), piezoelectric (strain), triboelectric (motion), flexoelectric (strain gradient) and photovoltaic (solar excitation) charge generation mechanisms that are used to control electrochemical reactions. (Reproduced with permission from Y. Zhang et al., *Chem. Soc. Rev.* 46, 2017, 7757–7786.)

10.2 SUPERCAPACITOR-BATTERY HYBRID DEVICES

Supercapacitors (SCs) and batteries share similar configurations with cathode and anode with active material, separator, electrolyte and current collector. However, their electrochemical characteristics make them two different devices. Hybridisation of these two energy storage devices depends not only on the micro-/nanostructure design of electrode materials but more critically depends on the configuration engineering of the device [5]. The hybridisation of the two devices facilitates the direct integration of high power and long lifetime from SCs as well as the high energy from batteries. Generally, battery-SC hybrid systems employ high-energy/sluggish redox electrodes on one side and low-energy/fast double-layer electrodes on the other side, which generate a larger working voltage together with higher capacitance.

The best performances of electrical double-layer capacitors (EDLCs) generally include good cyclic stability and high power densities, and the cyclic voltammetry (CV) is rectangular in shape (Figure 10.2) [6]. In case of rectangular CV or linear V-shaped charge-discharge (CD) in EDLCs, instantaneous charge separation due to polarisation occurs under the external electric field, and as a result dV/dt is constant. In the case of batteries, a distinct peak in the CV and plateaus in the CD are witnessed due to phase transformation of the material. In contrast, pseudocapacitive material does not show any phase transformation; instead, a highly reversible change in the oxidation state appears (Figure 10.2b and c). The charge storage mechanism is different in hybrid devices, where one electrode exploits the double-layer storage mechanism, whereas the other electrode stores charge by means of faradaic reactions. Therefore, the concept of an asymmetric hybrid cell allowed us to produce an ideal combination of positive and negative electrode, so that it can fulfil the criteria of smart electronics.

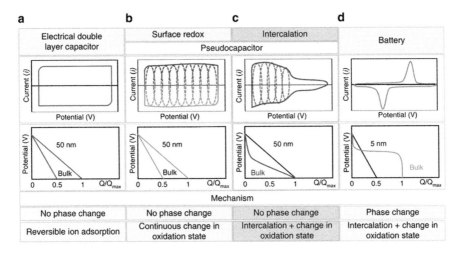

FIGURE 10.2 Faradic and capacitive energy storage. (a) Double-layer capacitor, (b) surface redox pseudocapacitance due to adsorption and/or fast intercalation of ions, (c) intercalation pseudocapacitance and (d) batteries. (i = current, v = sweep rate.) (Reproduced with permission from M.R. Lukatskaya et al., *Nat. Commun.* 7, 2016, 12647.)

10.2.1 LITHIUM-ION CAPACITOR

Li-ion battery material, LiFePO$_4$ is the most studied cathode material in lithium-ion capacitors (LICs). Graphene-based electrode materials used in both anode and cathode have the potential to improve both energy and power density performance (Figure 10.3). Recently, a Fe$_3$O$_4$ nanoparticle/graphene (Fe$_3$O$_4$/G) composite was prepared for the negative electrode and three-dimensional (3D)-graphene as positive electrode was tested for a hybrid SC with an improved ultrahigh energy density of 147 W h kg^{-1} (power density of 150 W kg^{-1}), which also remains at 86 W h kg^{-1} even at a high power density of 2587 W kg^{-1} [7]. Naoi et al. have synthesised a nanostructured core (LiFePO$_4$)/shell (graphitic carbon) material that shows ultrafast charge-discharge performance (beyond the 300 C rate) for designing high-energy and high-power hybrid SCs [8]. Another report by Salvatierra et al. used a scalable method for the growth of graphene-CNT heterostructures from multidimensional carbon substrates as anodes and cathodes for binder-free LICs [9]. The hybrid device has shown high-energy densities (~120 W h kg^{-1}) and high-power density (~20,500 W kg^{-1} at 29 W h kg^{-1}) with a large operating voltage window (4.3 V). Aqueous electrochemical flow capacitors have been demonstrated by Liu et al. [10]. In that paper, LiMn$_2$O$_4$ and activated carbon slurry were employed as electrodes. A record high energy density of 23.4 W h kg^1 at a power density of 50 W kg^1 has been achieved for aqueous flow capacitors tested at static conditions.

Another new ultrafast Li$_4$Ti$_5$O$_{12}$ (LTO) nanocrystal electrode material has been explored recently for capacitive energy storage in a hybrid device, called a *nanohybrid capacitor*, named by Naoi et al. [11]. Most of the nanohybrid capacitors deal with the carbon nanofiber (CNF), graphene, biomass-derived carbon and activated carbon (AC) as cathode and LTO/CNF as anode materials [11–25].

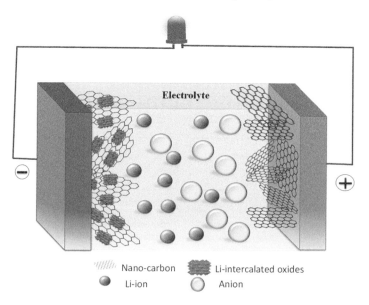

FIGURE 10.3 A Li-ion capacitor.

10.2.2 Sodium-Ion Capacitor

Graphene, transition metal oxide/graphene composites and biomass-derived carbon have been proven as potential candidates for sodium-ion (Na-ion) hybrid SCs. Nanocomposites comprising Nb_2O_5@carbon core–shell nanoparticles (Nb_2O_5@C) and reduced graphene oxide were utilised for hybrid Na-ion capacitors (NICs). Ramakrishnan et al. recently developed a synthesised biomass (goat hair)–derived activated carbon cathode with a high surface area of 2042 $m^2\ g^{-1}$ and MoO_2@rGO composite anode materials for hybrid NICs [26]. Ding et al. have reported a NIC with peanut shell–derived carbon material [27]. They utilised the unique structure of the peanut shell and achieved two fundamentally different (anode versus cathode) very high performance electrode materials from the same precursor with a long cycle life. Wang et al. demonstrated that a graphene combination with Ti_3C_2 Mxene can be used to derive TiO_2 mesostructure-rGO nanocomposites that are promising candidates for the NIC anode [28].

10.3 SUPERCAPACITOR-NANOGENERATOR HYBRID TECHNOLOGY

Researchers are working on integrating the nanogenerator with EES for utilising the generated electricity from the nanogenerator to the storage part of the device for portable electronic devices, smartphones, bendable displays, wearable devices and medical devices. The all-in-one integration of an energy storage device with a nanogenerator is the eventual purpose for the development of the hybrid technology and device, which can be a possible solution for energy technology today [29]. Different energy harvesting technologies with nanogenerators can be put into categories based on their energy sources, such as mechanical and thermal energy, including piezoelectric nanogenerator (PENG), triboelectric nanogenerator (TENG) and pyroelectric nanogenerator (PYNG). The performance of these nanogenerators and their hybridisation with EES in a single cell are discussed here. The roles of carbonaceous material such as graphene, CNTs, carbide-derived carbon and its composites have been explored as electrode materials for these hybridised devices due to low cost, flexibility, light weight and high output of the materials.

The mechanical energy, stress or strain, can be transformed into electrical energy using the piezoelectric influence. The origin of stress/strain exists everywhere, in vibrations, acoustic noise, body motion and airflow. The first attempt to fabricate a piezoelectric nanogenerator–driven self-charging SC was based on polyvinylidene difluoride (PVDF)-ZnO as a piezoelectric as well as a separator, poly(vinylalcohol)-phosphoric acid (PVA- H_3PO_4) as a gel electrolyte, and electrochemically active manganese oxide (MnO_2) nanowires as positive and negative electrodes. Figure 10.4 gives the schematic representation of the hybrid energy system [30]. The working principle of the system is based on storing energy that was harvested by piezoelectric-driven electrochemical oxidation and reduction reaction. With the hybrid device, we seek to improve the integration level and reduce redundant energy loss in the power-management circuits between nanogenerator and energy storage devices. Song et al. developed a piezo-supercapacitor with polarised PVDF films coated with H_2SO_4/PVA

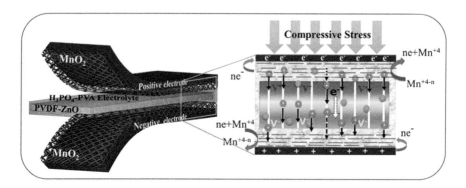

FIGURE 10.4 The hybrid piezoelectric nanogenerator–supercapacitor device and the charge storage mechanism. (Reproduced with permission from A. Ramadoss et al., ACS Nano. 9, 2015, 4337–4345.)

gel electrolyte as piezoelectric material and functionalised carbon cloths assembled with H_2SO_4/PVA electrolyte as both anode and cathode as SC [31]. Recently, Wang et al. developed a new nano-energy cell (NEC) that has high-density piezoelectric nanowires to harvest mechanical energy and a large electrolyte (H_3PO_4/PVA gel electrolyte)-nanowire interface to store electricity in the all-in-one system consisting of a PENG and SC [32].

TENGs provide a new approach to harvesting electricity from mechanical energy to operate small electronic devices. The working mechanism of TENGs was first explained on the basis of electrostatic and contact electrification physics by Wang and co-workers [32]. Carbonaceous materials have drawn huge interest in the development of a self-charging SC from a TENG. A flexible self-charging nanogenerator was developed based on folded carbon paper for harvesting mechanical energy from human motion and power portable electronics [33]. In this study, the device consists of a TENG and an SC, based on folded carbon paper, as energy harvester and storage device. TENG has been proven to be competent in harvesting the mechanical energy of human movements at high efficiency and large output power density. Recently, researchers have begun to integrate TENGs with SCs or LIBs using a rectifier. Recently, a self-powered electronic watch has been developed with a hybridised electromagnetic (EMG)-TENG for harvesting biomechanical energy from the natural motions of the wearer's wrist to sustainably power that energy to the storage device [34]. The mechanism of the hybrid nanogenerator is to utilise the collision between a magnetic ball and the coils to stimulate the instantaneous working of TENG and EMGs. A wearable self-charging power system that combines energy-harvesting and energy-storing (SCs) technologies based on a conformal nickel layer and rGO film coated on the surface of common polyester yarns were demonstrated by Pu et al. [35]. Guo et al. described an all-in-one shape-adaptive self-charging power generator based on a TENG and EDLC that has been simultaneously demonstrated for harvesting body motion energy to sustainably drive wearable/portable electronics [36].

PYNGs also have the potential to harvest temperature changes associated with exothermic or endothermic reactions [37]. Zhao et al. fabricated a flexible pyroelectric energy harvester device, composed of a CNT/PVDF/CNT sandwich as a potential approach for scavenging heat from chemical processes. A flexible hybrid nanogenerator that is capable of simultaneously or individually harvesting ambient thermal and mechanical energies was also developed with polyvinylidene difluoride (PVDF) as the active pyroelectric material and a patterned PTFE/Al combination as the TENG [38]. This device used a rectifier to maintain constant polarity at the steel cathode and carbon anode.

10.4 HYBRID DEVICE OF SOLAR CELL WITH ENERGY STORAGE SYSTEM

Electrochemical systems including batteries and electrochemical supercapacitors designed for energy storage can be coupled with a solar cell to not only store the charges but also balance the solar electricity fluctuations by acting as both energy storage and output regulator [39]. In 2011, Bae et al. showed that twisted fiber-like electrodes wrapped with ZnO NWs and graphenes can harvest solar energy using the dye-sensitized solar cells (DSSCs) and store it [40]. Recently, researchers attempted to focus on the DSSCs or organic solar cells (OSCs) as the energy source with an SC or battery for energy storage in either a planar structure or a fiber shape, which have shown their ability as self-charging and storage units. Liu et al. reported a self-charging power unit that combines a hybrid silicon nanowire-polymer heterojunction solar cell with a polypyrrole-based SC to simultaneously harvest solar energy and store it within the device [39]. Figure 10.5 gives a schematic representation of the same system. The hybrid device is able to generate photoelectric conversion to storage efficiency of 10.5%, which is proof of a highly efficient integrated hybrid device of solar energy conversion and storage. Another flexible printable DSSC-SC integrated energy device has been designed by Dong et al., which is capable of charging the device with potentials up to

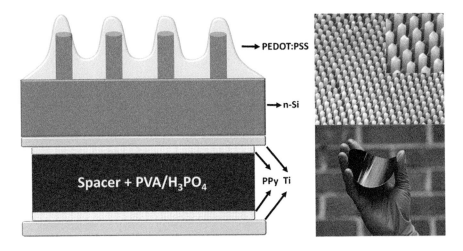

FIGURE 10.5 The integrated hybrid device containing a SiNW-based heterojunction solar cell and a polypyrrole-based supercapacitor. Nanowire array and optical image of the device. (Reproduced with permission from R. Liu et al., *Nano Lett.* 17, 2017, 4240–4247.)

1.8 V [41]. In this study, an SC was assembled using reduced graphene oxide (rGO) as the active material on the electrodes and a solid polymer electrolyte. Du et al. developed a hybrid device with self-stacked solvated graphene films as the free-standing electrode for flexible solid-state SCs, which is integrated with high-performance perovskite hybrid solar cells (pero-HSCs) to build self-powered electronics [42].

A new approach for integrating solar cells and batteries using a tri-iodide/iodide redox shuttle to couple a built-in dye-sensitized titanium dioxide photoelectrode with an oxygen electrode for the photo-assisted charging of a lithium-oxygen battery was recently developed by Yu et al. [43]. This concept of a 'photoassisted charging process' provides a new strategy to address the overpotential issue of non-aqueous Li-O_2 batteries and also provides a novel approach for integrating solar cells and batteries for an efficient hybrid device.

10.5 SUPERCAPACITOR-BIOFUEL CELL HYBRID TECHNOLOGY

Biofuel cells (BFCs) that convert chemical energy into electrical energy by the catalytic reaction of enzymes have become of interest as a research topic in the last few years and have shown great potential as renewable, efficient, power-generating devices for implantable applications. The smart integration of BFCs with energy storage devices, such as SCs and batteries was achieved in order to harvest significantly high power output [44]. The hybrid devices of BFCs with SCs and batteries are sometimes referred to as bio-capacitors and bio-batteries, respectively. Carbonaceous materials such as activated carbon, carbon nanotubes and graphene are mostly used as electrode material for hybrid devices.

Recently, a hybrid system with a CNT matrix as an SC and a CNT/enzyme matrix as electrode material for a biofuel cell setup was demonstrated as shown in Figure 10.6 [45]. The SC-BFC hybrid system provided high power discharge cycles where the CNT matrix was continuously charged through the biocatalytic energy conversion. Another self-charging electrochemical bio-capacitor was reported with CNT/polyaniline composites with interconnecting CtCDH/AuNPs/GF bioanode and a MvBOx/AuNPs/GF biocathode [46]. Knoche et al. reported an SC/biobattery

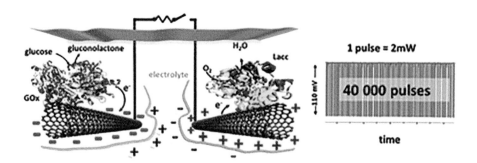

FIGURE 10.6 The supercapacitor/biofuel cell hybrid system. (Reproduced with permission from C. Agnès et al., *Energy Environ. Sci.*, 7, 2014, 1884–1888.)

hybrid system, where an oxygen-reducing cathode of carbon nanotubes modified with bilirubin oxidase immobilised with anthracene and tetrabutylammonium bromide-modified Nafion is coupled with a glucose bioanode of flavin adenine dinucleotide-dependent glucose dehydrogenase [47]. This device results in a specific capacitance of 300 F g^{-1}, which is significantly higher than other bio-capacitors. Therefore, practical applications of hybrid bio-devices are dependent on the biofuels used, the biocatalysts and the design of the device. The hybrid device of self-charging bio-supercapacitors/charge-storing BFCs can be recycled for short-time, low-current applications, functioning in solutions containing carbohydrates, alcohols or other energy-rich compounds.

10.6 HYBRID SELF-POWERED WATER ELECTROLYSIS TECHNOLOGIES

In this section we discuss recent progress in the field of energy conversion and generation, focussing on the self-powered electro- and photocatalysts and their control and processing using energy harvesting materials and devices. Energy harvesting deals with the conversion of ambient forms of energy such as light, heat, mechanical motion and wind which are wasted until harvested and if not harvested efficiently. A nanogenerator is one such energy harvester that harvests motion, heat or light into electrical energy. Solar cell and metal air batteries are also candidates among the energy harvesting units. The harvested energy is either utilised directly or rectified and stored in an SC or battery for application in a self-powered system (Figure 10.7). Here we

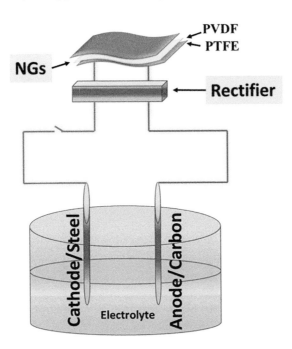

FIGURE 10.7 Self-powered water-splitting device integrated with nanogenerator.

discuss the coupling of energy harvesting devices and materials with electro-chemical (photochemical) systems. Basically we discuss energy harvesting units coupling with electrochemical systems, and the same can be considered in photocatalysis systems (solar cell as a harvesting unit).

Nanogenerator classes include pyroelectric [48], piezoelectric [49], triboelectric [50], thermoelectric [51], flexoelectric [52] and photovoltaics [53]. The piezoelectric, pyroelectric and flexoelectric classes deal with the generation of electric charge as a result of the change of applied stress, temperature change and strain gradient, respectively, and the photovoltaic deals with the generation of electron transfer to the conduction band to generate electric current. Recent requirements of countries demand the removal of fossil fuels and lead towards sustainable generation of energy. Hybridising the electrochemical or photoelectrochemical devices with self-powered systems provides an efficient way to meet current demands in the field of energy. Yan Zhang and co-workers successfully demonstrated a hybrid nanogenerator water-splitting device [4].

Recently, solar cell water-splitting hybridisation was reported by various research groups because of the growing sustainable energy demands. There are many reports that are available on solar cell water splitting [54]. In a solar cell–water splitting hybrid, sufficient voltage is provided to split the water into oxygen and hydrogen, and that voltage is provided from the self-powered solar cell (Figure 10.8), which is renewable in nature and focussed towards sustainable development.

Metal air batteries are self-charging electrochemical cells that have high energy density. This type of electrochemical cell uses anodes that are fabricated from pure metal and an external cathode of ambient air. The hybridisation of water splitting with metal air batteries has also been performed, and a lot of research is going on in this particular area [54]. The concept here is the same as that of hybridisation. The power sources utilised here are metal air batteries to split water.

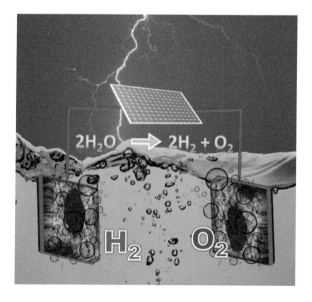

FIGURE 10.8 Self-powered water-splitting device integrated with solar cell.

10.7 HYBRID TECHNOLOGY WITH FUEL CELL AND ENERGY STORAGE DEVICES

Hybrid power sources, such as a fuel cell connected with another power supply, capacitors, on secondary batteries are the solution for the requirement of fuel cell output into the mean value in a short period of time by providing pulsed production from alternative power sources. Lee et al. in 2004 developed a direct methanol fuel cell (DMFC)/battery hybrid power system for portable applications, where the DMFC was applied for the main power source at average load and the battery was applied for auxiliary power at overload [55]. Recently, a hybrid system of fuel cell and electrochemical capacitor was developed that consisted of a 25 W proton exchange membrane fuel cell stack connected in parallel with a 70 F capacitor bank [56]. The hybrid system continuously powered the same regime for 25 hours, and the operating potential never reached the voltage cut-off point, not even during the high load of 18 W, as reported by Jarvis et al.

10.8 OTHER HYBRID TECHNOLOGIES

There is a huge demand among researchers for hybrid energy technologies including self-powered microdevices for the development of future smart electronics systems. In a recent report, Gu et al. presented a flexible micro-supercapacitors integrated with photodetectors to form a highly compact photo-detecting system [57]. Reduced graphene oxide/Fe_2O_3 hollow nanospheres were used for on-chip micro-supercapacitor fabrication, and CdS nanowire was used for the fabrication of the photodetectors.

Pankratova et al. demonstrated a new type of supercapacitive photo-bioanode and biosolar cell for simultaneous solar energy conversion and storage in the form of electric charge [58]. Kadimisetty et al. fabricated a low-cost, sensitive, supercapacitor-powered electrochemiluminescent (ECL) protein immunoarray for a cancer biomarker that employs screen-printed carbon sensors with gravity flow for sample/reagent delivery and washing [59].

Pan et al. developed a hybrid nanogenerator/biofuel cell made up of a fiber nanogenerater and a fiber biofuel cell onto a carbon fiber [60]. The nanogenerator was fabricated by etching the ZnO NW film at one end of the carbon fiber, contacting the top surface using silver paste and tape, and leading out two electrodes from the surface and the core electrodes. On the other side of the electrode, a layer of soft epoxy polymer was coated on the carbon fiber as an insulator, then two gold electrodes were patterned onto it and coated with CNTs/glucose oxidase and laccase to form the anode and cathode.

10.9 SUMMARY AND OUTLOOK

It is highly desirable to develop sustainable power systems with energy harvesting from multiple energy sources in the self-powered storage system. This chapter described innovative approaches towards developing various forms of energy harvesting as well as energy storage systems available with nano-carbon-based materials. Various forms of energy harvesting systems such as solar, thermal, chemical and mechanical energy and their integration with energy storage devices were demonstrated to meet the challenge of large-scale world energy requirements. Carbonaceous materials such as graphene, CNTs, activated carbon and their composites have tremendous potential to act as active material for the hybrid devices. However, there are still huge improvements to be made in the performances of the devices in order to develop innovative technologies for maintenance-free self-powered hybrid systems. This area of research is wide open and successful collaboration between scientists and engineers is necessary to develop next-generation hybrid technology and self-powered devices.

REFERENCES

1. J.H. Lee, J. Kim, T.Y. Kim, M.S. Al Hossain, S.W. Kim, J.H. Kim, All-in-one energy harvesting and storage devices, *J. Mater. Chem. A*. 4, 2016, 7983–7999. doi: 10.1039/c6ta01229a
2. D.P. Dubal, O. Ayyad, V. Ruiz, P. Gómez-Romero, Hybrid energy storage: The merging of battery and supercapacitor chemistries, *Chem. Soc. Rev.* 44, 2015, 1777–1790. doi: 10.1039/C4CS00266K
3. L. Kouchachvili, W. Yaïci, E. Entchev, Hybrid battery/supercapacitor energy storage system for the electric vehicles, *J. Power Sources* 374, 2018, 237–248. doi: 10.1016/j.jpowsour.2017.11.040
4. Y. Zhang, M. Xie, V. Adamaki, H. Khanbareh, C.R. Bowen, Control of electro-chemical processes using energy harvesting materials and devices, *Chem. Soc. Rev.* 46, 2017, 7757–7786. doi: 10.1039/C7CS00387K
5. J. Yan, Q. Wang, T. Wei, Z. Fan, Recent advances in design and fabrication of electrochemical supercapacitors with high energy densities, *Adv. Energy Mater.* 4, 2014, 1300816. doi: 10.1002/aenm.201300816
6. M.R. Lukatskaya, B. Dunn, Y. Gogotsi, Multidimensional materials and device architectures for future hybrid energy storage, *Nat. Commun.* 7, 2016, 12647. doi: 10.1038/ncomms12647
7. F. Zhang, T. Zhang, X. Yang, L. Zhang, K. Leng, Y. Huang, Y. Chen, A high-performance supercapacitor-battery hybrid energy storage device based on graphene-enhanced electrode materials with ultrahigh energy density, *Energy Environ. Sci.* 6, 2013, 1623–1632. doi: 10.1039/c3ee40509e
8. K. Naoi, K. Kisu, E. Iwama, S. Nakashima, Y. Sakai, Y. Orikasa, P. Leone et al., Ultrafast charge–discharge characteristics of a nanosized core–shell structured LiFePO4 material for hybrid supercapacitor applications, *Energy Environ. Sci.* 9, 2016, 2143–2151. doi: 10.1039/C6EE00829A
9. R.V. Salvatierra, D. Zakhidov, J. Sha, N.D. Kim, S.K. Lee, A.R.O. Raji, N. Zhao, J.M. Tour, Graphene carbon nanotube carpets grown using binary catalysts for high-performance lithium-ion capacitors, *ACS Nano*. 11, 2017, 2724–2733. doi: 10.1021/acsnano.6b07707
10. H. Liu, L. Liao, Y.C. Lu, Q. Li, High energy density aqueous li-ion flow capacitor, *Adv. Energy Mater.* 7, 2017. doi: 10.1002/aenm.201601248
11. Katsuhiko Naoi, Wako Naoi, Shintaro Aoyagi, Jun-ichi Miyamoto, Takeo Kamino, New generation 'nanohybrid supercapacitor', *Acc. Chem. Res.* 46, 2013, 1075–1083.
12. B. Zhao, R. Ran, M. Liu, Z. Shao, A comprehensive review of $Li_4Ti_5O_{12}$-based electrodes for lithium-ion batteries: The latest advancements and future perspectives, *Mater. Sci. Eng. R. Rep.* 98, 2015, 1–71.
13. J. Kim, J. Kim, Y. Lim, J. Lee, Y. Kim, Effect of carbon types on the electrochemical properties of negative electrodes for Li-ion capacitors, *J. Power Sources* 196, 2011, 10490–10495. doi: 10.1016/j.jpowsour.2011.08.081

14. A. Jain, V. Aravindan, S. Jayaraman, P.S. Kumar, R. Balasubramanian, S. Ramakrishna, S. Madhavi, M.P. Srinivasan, Activated carbons derived from coconut shells as high energy density cathode material for Li-ion capacitors, *Sci. Rep.* 3, 2013, 3002. doi: 10.1038/srep03002
15. V. Aravindan, W. Chuiling, S. Madhavi, High power lithium-ion hybrid electrochemical capacitors using spinel LiCrTiO4 as insertion electrode, *J. Mater. Chem.* 22, 2012, 16026. doi: 10.1039/c2jm32970k
16. W.J. Cao, J.P. Zheng, Li-ion capacitors with carbon cathode and hard carbon / stabilized lithium metal powder anode electrodes, *J. Power Sources* 213, 2012, 180–185. doi: 10.1016/j.jpowsour.2012.04.033
17. A. Vlad, N. Singh, J. Rolland, S. Melinte, P.M. Ajayan, J.-F. Gohy, Hybrid supercapacitor-battery materials for fast electrochemical charge storage, *Sci. Rep.* 4, 2014, 4315. doi: 10.1038/srep04315
18. T. Yi, S. Yang, Y. Xie, Recent advances of $Li_4Ti_5O_{12}$ as a promising next generation anode material for high power lithium-, *J. Mater. Chem. A Mater. Energy Sustain.* 3, 2015, 5750–5777. doi: 10.1039/C4TA06882C
19. T. Yi, L. Jiang, J. Shu, C. Yue, R. Zhu, H. Qiao, Recent development and application of Li 4 Ti 5 O 12 as anode material of lithium ion battery, *J. Phys. Chem. Solids* 71, 2010, 1236–1242. doi: 10.1016/j.jpcs.2010.05.001
20. K. Naoi, W. Naoi, S. Aoyagi, J. Miyamoto, T. Kamino, New generation "nanohybrid supercapacitor," *Acc. Chem. Res.* 46, 2013, 1075–1083. doi: 10.1021/ar200308h.
21. R. Satish, V. Aravindan, W. Chui, S. Madhavi, Carbon-coated $Li_3V_2(PO_4)_3$ as insertion type electrode for lithium-ion hybrid electrochemical capacitors: An evaluation of anode and cathodic performance, *J. Power Sources* 281, 2015, 310–317. doi: 10.1016/j.jpowsour.2015.01.190
22. K. Naoi, S. Ishimoto, Y. Isobe, S. Aoyagi, High-rate nano-crystalline Li4Ti5O12 attached on carbon nano-fibers for hybrid supercapacitors, *J. Power Sources* 195, 2010, 6250–6254. doi: 10.1016/j.jpowsour.2009.12.104
23. D. Mhamane, V. Aravindan, M. Kim, H. Kim, K.C. Roh, D. Ruan, S.H. Lee, M. Srinivasan, K. Kim, ultracapacitor and Li-ion hybrid capacitor, *J. Mater. Chem. A Mater. Energy Sustain.* 4, 2016, 5578–5591. doi: 10.1039/C6TA00868B
24. K. Naoi, 'Nanohybrid capacitor': The next generation electrochemical capacitors, *Fuel Cells* 10, 2010, 825–833. doi: 10.1002/fuce.201000041
25. H. Kim, K. Park, M. Cho, M. Kim, J. Hong, S.-K. Jung, K.C. Roh, K. Kang, High-performance hybrid supercapacitor based on graphene-wrapped $Li_4Ti_5O_{12}$ and activated carbon, *ChemElectroChem.* 1, 2014, 125–130. doi: 10.1002/celc.201300186
26. K. Ramakrishnan, C. Nithya, R. Karvembu, High-performance sodium ion capacitor based on MoO_2@rGO nanocomposite and goat hair derived carbon electrodes, *ACS Appl. Energy Mater.* 1, 2018, 841–850. doi: 10.1021/acsaem.7b00284
27. J. Ding, H. Wang, Z. Li, K. Cui, D. Karpuzov, X. Tan, A. Kohandehghan, D. Mitlin, Peanut shell hybrid sodium ion capacitor with extreme energy-power rivals lithium ion capacitors, *Energy Environ. Sci.* 8, 2015, 941–955. doi: 10.1039/C4EE02986K
28. R. Wang, S. Wang, Y. Zhang, D. Jin, X. Tao, L. Zhang, Graphene-coupled Ti_3C_2 MXenes-derived TiO_2 mesostructure: Promising sodium-ion capacitor anode with fast ion storage and long-term cycling, *J. Mater. Chem. A.* 6, 2018, 1017–1027. doi: 10.1039/C7TA09153B
29. J. Kim, J.H. Lee, J. Lee, Y. Yamauchi, C.H. Choi, J.H. Kim, Research update: Hybrid energy devices combining nanogenerators and energy storage systems for self-charging capability, *APL Mater.* 5, 2017. doi: 10.1063/1.4979718
30. A. Ramadoss, B. Saravanakumar, S.W. Lee, Y.-S. Kim, S.J. Kim, Z.L. Wang, Piezoelectric-driven self-charging supercapacitor power cell, *ACS Nano.* 9, 2015, 4337–4345. doi: 10.1021/acsnano.5b00759
31. R. Song, H. Jin, X. Li, L. Fei, Y. Zhao, H. Huang, H. Lai-Wa Chan, Y. Wang, Y. Chai,

A rectification-free piezo-supercapacitor with a polyvinylidene fluoride separator and functionalized carbon cloth electrodes, *J. Mater. Chem. A.* 3, 2015, 14963–14970. doi: 10.1039/C5TA03349G
32. F. Wang, C. Jiang, C. Tang, S. Bi, Q. Wang, D. Du, J. Song, High output nano-energy cell with piezoelectric nanogenerator and porous supercapacitor dual functions – A technique to provide sustaining power by harvesting intermittent mechanical energy from surroundings, *Nano Energy* 21, 2016, 209–216. doi: 10.1016/j.nanoen.2016.01.018
33. C. Zhou, Y. Yang, N. Sun, Z. Wen, P. Cheng, X. Xie, H. Shao et al., Flexible self-charging power units for portable electronics based on folded carbon paper, *Nano Res.* 2, 2018, 1–10. doi: 10.1007/s12274-018-2018-8
34. T. Quan, X. Wang, Z.L. Wang, Y. Yang, Hybridized electromagnetic-triboelectric nanogenerator for a self-powered electronic watch, *ACS Nano.* 9, 2015, 12301–12310. doi: 10.1021/acsnano.5b05598
35. X. Pu, L. Li, M. Liu, C. Jiang, C. Du, Z. Zhao, W. Hu, Z.L. Wang, Wearable self-charging power textile based on flexible yarn supercapacitors and fabric nanogenerators, *Adv. Mater.* 28, 2016, 98–105. doi: 10.1002/adma.201504403
36. H. Guo, M.H. Yeh, Y.C. Lai, Y. Zi, C. Wu, Z. Wen, C. Hu, Z.L. Wang, All-in-one shape-adaptive self-charging power package for wearable electronics, *ACS Nano.* 10, 2016, 10580–10588. doi: 10.1021/acsnano.6b06621
37. T. Zhao, W. Jiang, D. Niu, H. Liu, B. Chen, Y. Shi, L. Yin, B. Lu, Flexible pyroelectric device for scavenging thermal energy from chemical process and as self-powered temperature monitor, *Appl. Energy* 195, 2017, 754–760. doi: 10.1016/j.apenergy.2017.03.097
38. H. Zhang, S. Zhang, G. Yao, Z. Huang, Y. Xie, Y. Su, W. Yang, C. Zheng, Y. Lin, Simultaneously harvesting thermal and mechanical energies based on flexible hybrid nanogenerator for self-powered cathodic protection, *ACS Appl. Mater. Interfaces* 7, 2015, 28142–28147. doi: 10.1021/acsami.5b10923
39. R. Liu, J. Wang, T. Sun, M. Wang, C. Wu, H. Zou, T. Song et al., Silicon nanowire/polymer hybrid solar cell-supercapacitor: A self-charging power unit with a total efficiency of 10.5%, *Nano Lett.* 17, 2017, 4240–4247. doi: 10.1021/acs.nanolett.7b01154
40. J. Bae, Y.J. Park, M. Lee, S.N. Cha, Y.J. Choi, C.S. Lee, J.M. Kim, Z.L. Wang, Single-fiber-based hybridization of energy converters and storage units using graphene as electrodes, *Adv. Mater.* 23, 2011, 3446–3449. doi: 10.1002/adma.201101345
41. P. Dong, M.F. Rodrigues, J. Zhang, R.S. Borges, K. Kalaga, A.L.M. Reddy, G.G. Silva, P.M. Ajayan, J. Lou, A flexible solar cell/supercapacitor integrated energy device, *Nano Energy* 42, 2017, 181–186. doi: 10.1016/j.nanoen.2017.10.035
42. P. Du, X. Hu, C. Yi, H.C. Liu, P. Liu, H.L. Zhang, X. Gong, Self-powered electronics by integration of flexible solid-state graphene-based supercapacitors with high performance perovskite hybrid solar cells, *Adv. Funct. Mater.* 25, 2015, 2420–2427. doi: 10.1002/adfm.201500335
43. M. Yu, X. Ren, L. Ma, Y. Wu, Integrating a redox-coupled dye-sensitized photoelectrode into a lithium-oxygen battery for photo-assisted charging, *Nat. Commun.* 5, 2014, 1–6. doi: 10.1038/ncomms6111
44. D. Pankratov, Z. Blum, S. Shleev, Hybrid electric power biodevices, *ChemElectroChem.* 1, 2014, 1798–1807. doi: 10.1002/celc.201402158
45. C. Agnès, M. Holzinger, A. Le Goff, B. Reuillard, K. Elouarzaki, S. Tingry, S. Cosnier, Supercapacitor/biofuel cell hybrids based on wired enzymes on carbon nanotube matrices: Autonomous reloading after high power pulses in neutral buffered glucose solutions, *Energy Environ. Sci.* 7, 2014, 1884. doi: 10.1039/c3ee43986k
46. D. Pankratov, Z. Blum, D.B. Suyatin, V.O. Popov, S. Shleev, Self-charging electrochemical biocapacitor, *ChemElectroChem.* 1, 2014, 343–346. doi: 10.1002/celc.201300142

47. K.L. Knoche, D.P. Hickey, R.D. Milton, C.L. Curchoe, S.D. Minteer, Hybrid glucose/O_2 biobattery and supercapacitor utilizing a pseudocapacitive dimethylferrocene redox polymer at the bioanode, *ACS Energy Lett.* 1, 2016, 380–385. doi: 10.1021/acsenergylett.6b00225
48. C.R. Bowen, J. Taylor, E. Leboulbar, D. Zabek, A. Chauhan, R. Vaish, Pyroelectric materials and devices for energy harvesting applications, *Energy Environ. Sci.* 7, 2014, 3836–3856. doi: 10.1039/c4ee01759e
49. C.R. Bowen, H.A. Kim, P.M. Weaver, S. Dunn, Piezoelectric and ferroelectric materials and structures for energy harvesting applications, *Energy Environ. Sci.* 7, 2014, 25–44. doi: 10.1039/c3ee42454e
50. Z.L. Wang, On Maxwell's displacement current for energy and sensors: The origin of nanogenerators, *Mater. Today* 20, 2017, 74–82. doi: 10.1016/j.mattod.2016.12.001
51. G. Tan, L.-D. Zhao, M.G. Kanatzidis, Rationally designing high-performance bulk thermoelectric materials, *Chem. Rev.* 116, 2016, 12123–12149. doi: 10.1021/acs.chemrev.6b00255
52. Z.L. Wang, Triboelectric nanogenerators as new energy technology and self-powered sensors – Principles, problems and perspectives, *Faraday Discuss.* 176, 2014, 447–458. doi: 10.1039/C4FD00159A
53. P. Zubko, G. Catalan, A.K. Tagantsev, Flexoelectric effect in solids, *Annu. Rev. Mater. Res.* 43, 2013, 387–421. doi: 10.1146/annurev-matsci-071312-121634
54. L. Wei, K. Goh, Ö. Birer, H.E. Karahan, J. Chang, S. Zhai, X. Chen, Y. Chen, A hierarchically porous nickel–copper phosphide nano-foam for efficient electrochemical splitting of water, *Nanoscale.* 9, 2017, 4401–4408. doi: 10.1039/C6NR09864A
55. B. Do Lee, D.H. Jung, Y.H. Ko, Analysis of DMFC/battery hybrid power system for portable applications, *J. Power Sources* 131, 2004, 207–212. doi: 10.1016/j.jpowsour.2003.12.063
56. L.P. Jarvis, L.P. Jarvis, T.B. Atwater, T.B. Atwater, P.J. Cygan, P.J. Cygan, Fuel cell electrochemical capacitor hybrid for intermittent high power applications, *J. Power Sources* 79, 1999, 60–63. doi: 10.1016/S0378-7753(98)00199-2
57. S. Gu, Z. Lou, L. Li, Z. Chen, X. Ma, G. Shen, Fabrication of flexible reduced graphene oxide/Fe_2O_3 hollow nanospheres based on-chip micro-supercapacitors for integrated photodetecting applications, *Nano Res.* 9, 2016, 424–434. doi: 10.1007/s12274-015-0923-7
58. G. Pankratova, D. Pankratov, K. Hasan, H.E. Åkerlund, P.Å. Albertsson, D. Leech, S. Shleev, L. Gorton, Supercapacitive photo-bioanodes and biosolar cells: A novel approach for solar energy harnessing, *Adv. Energy Mater.* 7, 2017, 1–5. doi: 10.1002/aenm.201602285
59. K. Kadimisetty, I.M. Mosa, S. Malla, J.E. Satterwhite-warden, T.M. Kuhns, R.C. Faria, N.H. Lee, J.F. Rusling, Biosensors and bioelectronics immunoarray, *Biosens. Bioelectron.* 77, 2015, 188–193. doi: 10.1016/j.bios.2015.09.017
60. C. Pan, Z. Li, W. Guo, J. Zhu, Z.L. Wang, Fiber-based hybrid nanogenerators for/as self-powered systems in biological liquid, *Angew. Chem., Int. Ed.* 50, 2011, 11192–11196. doi: 10.1002/anie.201104197

11 State-of-the-Art Renewable Energy Technology

11.1 INTRODUCTION

Affordable and sustainable clean energy technology is the greatest challenge for society in the twenty-first century [1]. Enhanced living standards of developed countries and their growing dependence on technology as well as increasing populations of the developing countries certainly have made energy the most essential commodity to smoothly run our lives as well as our economies. As the non-renewable energy resources, for example fossil fuels, reserves are limited and they continue to increasingly contribute to global CO_2 emissions, to keep our earth safer and counter potential environmental threats, several renewable energy sectors based on self-sustainable, pollution-free and affordable technologies are revolutionizing globally [2–6]. Renewable energies can be harvested in a decentralized manner to meet rural and small-scale energy requirements and therefore carry a beacon of hope for providing electricity to the remotest corners of the earth. Many developments in the past few years in the state-of-the-art materials and technologies are responsible for advancement of renewable energy. Currently, renewable energy sources supply about 23.7% of total world energy demand [7], as compared to only 2% in 1998 [8]. Convincing evolution occurred in the renewable energy sector mainly in the field of solar photovoltaic (PV), wind power and hybrid storage-based renewable energy, while the growth in other sectors is quite slow in terms of power generation and storage [2,6,9].

There are a number of ways to generate green and renewable energy, for example, wind, solar, tidal, biomass and geothermal sources. The natural potential for renewable energy is enormous. For example, covering just a small region of Arizona with state-of-the-art PV cells would produce enough electricity for the entire United States [10], and installing modern wind turbines around the North Sea region may provide enough electricity for the entire European Union population [11]. However, almost all of these sources are intermittent and dispersed as compared to the isolated large-scale facilities that currently supply the majority of electrical energy. In other words, they depend on weather conditions, geometrical locations of the fields and opposite integration into the grid system. Therefore, to make the best use of these energy sources and their appropriate distribution to end-users in a stable and controlled manner, efficient energy storage systems are also required. Recently, researchers have been working on the concept of sustainability of storage via electrochemical as well as thermal routes, although challenges remain [2,5]. Additionally, to achieve the best output, engineering of the materials and design of the devices also play significant roles. In the past few decades, researchers took advantage of the physical and chemical properties of a myriad of carbonaceous materials in the form of graphene, carbon nanotubes, activated carbon and fullerene and used them as active materials in the different state-of-the-art sustainable energy technologies.

11.2 VARIOUS RENEWABLE ENERGY TECHNOLOGIES

Five of the most studied renewable energy technologies such as storage, solar, wind, water and waste resources are discussed in this chapter.

11.2.1 State-of-the-Art Storage Technologies

In the search for sustainable alternatives to decaying fossil fuels and intermittent renewable energies, various energy storage technologies are now being considered as a valuable link in the modern electricity value chain. Among a myriad of energy storage systems, such as pumped hydro storage, compressed air energy storage and flywheel energy storage, electrochemical energy storage (EES) systems are fast gaining in popularity for an energy-efficient future [12–14]. The two most contemporary EES subclasses are rechargeable batteries (such as lithium-ion batteries [LIBs]) and supercapacitors (SCs), which have taken the commercial storage market by storm. Although batteries and SCs largely differ on the basis of their charge-storage mechanism [15], their storage applications cover sectors such as hybrid electric vehicles (HEVs), portable electronics and wireless network communication systems as well as military marine and aerospace missions simultaneously depending on the energy density or power density needed.

In terms of technological maturity, operational battery storage technologies, such as LIBs, are much more developed as compared to SCs because of the ideal energy density that is offered by them ($\sim$200 W h Kg^{-1}) [16]. The most common form of operational battery storage technology is LIB followed by Li-ion phosphate battery, sodium sulfur (NaS) battery, vanadium redox flow and lead-acid battery, respectively. LIBs mostly dominate the portable energy storage applications, contributing to about 18% of the total battery energy storage systems, while NaS contributes to at least 24% indicating their importance in large-scale storage purposes [17]. A recent Frost and Sullivan report demonstrates the growing popularity of LIBs and says that between 2012 and 2026, lithium industries' growth almost doubled from US$11.7 billion to US$22.5 billion [18]. Some leading LIB manufacturing companies are the independent subsidiary units of electronic enterprises, such as Samsung, Panasonic, Toshiba, LG Chem, etc. as well as leading automotive ventures, like Tesla.

Although SC technology is developed and is commercially available, it is mainly used in demonstration projects. SCs, especially electrochemical double-layer capacitors (EDLCs), are extremely useful for diverse applications such as uninterruptible power supply in electric grids as well as regenerative power brakes and AC line-filtering. The SC market is expected to flourish and top US$11 billion by 2024 because of the many technical advantages of SCs, including charging the systems in seconds with high power delivery (>10 KW Kg^{-1}) [19] and stable cycle life ($\sim 10^5$ continuous charge-discharge cycles), although low energy density and a high cost-to-performance ratio become the bottlenecks of this industry's growth [20]. Some modifications in terms of material development (introduction of redox species to carbon matrices to generate pseudocapacitance) [21,22] and device architecture (ultrathin planar micro-supercapacitors) [23,24] as well as electrolyte selection (organic, ionic liquids to enhance potential window) [25] are promising technological improvements that will enhance the operating potential window and volumetric energy density of conventional SCs. Maxwell Technologies, Cellergy, Nippon Chemi-Con and Paper Battery Company are some of the emerging manufacturers of SCs that deal with heavy power supply applications of the technology as well as produce ultrathin SCs as a replacement to conventional LIBs for applications in consumer electronics and wireless sensors.

A hybrid SC, such as a lithium-ion hybrid supercapacitor (LIHS) [26–28] with a capacitive electrode as a source of power and a faradic electrode as a source of energy may offer a promising solution to super-fast charging as well as efficient storage for our personal electronic gadgets like mobile phones or even electric cars. But this type of technology is still very new and needs more research and monetary investment.

Another current research scenario in this field offers conjugation of energy storage systems with energy harvesting units like solar cells or nanogenerators for the development of self-powered electrochemical energy storage systems (SEESs) [29–31] that harvest their operating energy from the environment, for example solar energy [32], or from small-scale mechanical energy sources, like bodily movements [33].

11.2.2 SOLAR ENERGY TECHNOLOGY

Solar energy technology is perhaps the most reliable and fastest-growing renewable technology for sustainable electricity generation. Solar energy can be harnessed by both PV conversion as well as thermal conversion. In 1954, the first silicon 'photocell for converting solar radiation into electrical power' was reported with an efficiency of 6%, and that led to a sudden escalation in the development of PV cell technology [34]. In the last decade, PV cell technology has stepped up from an outsider status to one of the strongest contributors to the ongoing energy transition, with a whopping 1.7% contribution to world solar cell electricity production from a negligible 0.01% contribution in 1990 [35]. According to the International Energy Agency (IEA), PV technology is the fastest-growing technology and was expected to pass the 300 GW global installation in 2017 [30]. Three generations of PV solar cells exist in the market:

a. *Silicon-based solar cells (mono- and polycrystalline silicon)*: The first-generation solar PV cells, mainly aimed at microelectronics, that make up 80% of the global coverage and currently represent 90% of the market share.
b. *Thin-film solar cells (CdTe, copper indium gallium selenide [CIGS] and amorphous Si)*: The second-generation PV solar cells as a cheaper alternative to crystalline Si solar cells. Although they suffer from lower efficiencies than first-generation solar cells, they have better mechanical properties and provided opportunity in new fields like electrochemistry.

c. *Tandem, Perovskite, dye-sensitized solar cells (DSSCs), organic and hybrid solar cells*: The third-generation solar cells are aimed at broad concepts and applications, ranging from low-cost, low-efficiency systems (DSSC, organic) to high-cost, high-efficiency systems (III–V multi-junctions) for building integration to space applications, such as artificial satellites. They are still referred to as 'emerging concepts' because of their relatively low penetration in the commercial solar PV cell market.

Over the years, the technology has been substantially improved and consequently, production costs decreased and the emission of greenhouse gases decreased. For example, the trend is that with each doubling of the installed capacity of the mono- or polycrystalline silicon solar cells, the cumulative energy demand was reduced to at least 12% with a reduction of selling price by 20% as well as a decrease in greenhouse emissions by 20% [36]. Solar PV cells have a remarkable shelf-life of 25–40 years; therefore, it helps to produce around 14–20 times more energy in their lifetime than the energy invested in the production of the device. In 2017, a total of 99.1 GW grid-connected solar panels were installed as compared to 76.6 GW in 2016, a whopping 30% year-on-growth with China leading the global solar market demand, installing more than half of the world's solar capacity in a year, to be exact 53.3%. It has been a remarkable year for India too, with a staggering 127% solar market growth (9.6 GW from 4.3 GW in 2016). Under optimal conditions, the world's solar generation plant capacity will reach 1270.5 GW by 2022, meaning solar energy would reach TW production capacity level by 2022 [37]. However, challenges still exist in terms of achieving closer to thermodynamic efficiency, developing cheaper materials with excellent mechanical properties as well as recycling the device with minimal toxicity; these issues still need to be addressed.

Thermal conversion of solar energy to electricity is another bright energy generation prospect of the wide solar energy spectrum [6,38]. In tropical countries such as India, Sri Lanka or Brazil, where solar energy is an abundant source throughout the year, the solar thermal conversion process offers a cheap and sustainable electricity supply to even remote areas.

11.2.3 Wind Energy

Wind turbines convert the kinetic energy in wind to mechanical energy and then a generator converts it to electricity. Wind turbine systems have experienced huge substantial transformations over the years. In the 1980s, a 50 KW wind turbine was considered large, while a 2–3 MW is today's typical wind turbine output [39]. The installed capacity of wind power has increased from 4.8 MW in 1995 to more than 239 GW in 2011. It has made a significant contribution to the overall energy supply chains in China, the United States and Germany, with cumulative installed capacities of 62, 47 and, 29 GW, respectively, as of 2012 [40]. The IEA also estimated the global capacity by wind energy will even increase to almost 1282 TW h by 2020, a nearly 371% increase from 2009. By 2030, it will reach 2182 TW h, almost double from 2020, 80% of which will be contributed by off-shore wind turbines [40–41]. An example of technological achievement is represented by the completion and installation in early 2014 of the first Vestas V164, the world's most powerful wind turbine. Another example of an off-shore wind farm in Denmark includes Anholt 400 MW that is operated by Dong Energy and comprises 111 Siemens wind turbines, each rated 3.6 MW [42–43]. This shows the huge potential wind energy has as a renewable alternative, and developed countries such as the Netherlands, Denmark, etc. are harnessing most of their energy supply (∼60%) from this source.

11.2.4 WATER SPLITTING

Water can split into hydrogen and oxygen, but not at ambient conditions, which means the process is not a thermodynamically allowed process. It needs high temperature and pressure for water to split into its constituent molecules or the interference of some external energy sources, such as electricity (electrochemical water splitting) or light (photochemical water splitting) or both (photoelectrochemical water splitting) on catalyst materials that can effectively lower the activation energy of the process to help it occur at ambient conditions. Electrochemical or photochemical or photoelectrochemical water splitting is a long-term energy-conversion technology pathway that has the potential for low or almost no greenhouse gas emission. These technologies offer high conversion efficiency of a valued fuel like hydrogen even at a low operating temperature with the use of cheap transition-metal-based electrocatalysts or thin-film and/or semiconductor materials as photocatalysts. This energy technology is still in its nascent stage, and research efforts are ongoing in terms of material development, improvements in faradic or conversion efficiency, durability and cost of the energy conversion device.

11.2.5 WASTE TO ENERGY

Biomass burning has been one of the oldest practicing renewable technologies for heat and power production. Particularly in forest and sugar industries, solid biowastes – mainly wood chips and pellets – are used to provide process heat on site, with surplus electricity sold off-site to generate revenues [7]. The combustion of wet biomass wastes, such as sewage sludge and agricultural slurries in grate-firing boilers, often produces biogas (e.g. cow dung to methane) or other fuels to be used for heat generation and various other solid by-products that find use in various industries [4].

More recently, biowaste resources are being extensively studied because they are abundant, need to be recycled and are ample sources of carbon. The management of biowastes has always been a concern in smart cities; therefore, their conversion to economically worthwhile products for emergent applications has proven to be a greater motivated research focus than its erstwhile research orientation. Carbonization and activation of these biowaste materials mainly lead to the synthesis of activated porous carbons that have sufficiently high surface area, porosity as well as electrical conductivity. This promotes them as potential candidates to be employed as active electrode materials for electrochemical energy storage and conversion applications. Recently, a number of research papers have been published in this direction for the synthesis of carbonaceous materials for SC applications from plant leaves [44,45], coconut shell [46], biochar [47–49], waste corn shell [50], dry pine cones [51], fungus [52,53], eggshell [54] and even human hair [55,56]. Controlled synthesis engineering and use of appropriate biowaste precursors often produces selective carbonaceous structures, like few-layer-graphene [57,58] or carbon nanotubes [59,60] that not only have emergent energy applications but also reduce the production costs of these valuable products. Recently, few-layer-graphene produced from waste peanut shells [57] via a simple mechanical exfoliation technique has opened up a new path for industrial-scale cost-effective synthesis that may be used for storage purposes. Using biowaste for energy applications shows great promise, but this is still in the demonstration and experimentation stage and should be moved to the commercialization stage soon.

11.3 CONCLUSION

Renewable energy technologies are considered as one of the most important solutions for future energy demands, and they need to be developed in this century in order to replace the conventional fossil fuel–based energy technologies and reduce the global carbon footprint. Many emerging technologies exist, and they have varied levels of maturity. The scale of implementation is not even, and they are mostly dependent on weather conditions and are therefore difficult to integrate into centralized electric grid systems. For full penetration of the regenerative energies into the modern sustainable electric supply chain, smart grid systems with microgrids, energy storage units, a combination of electric power units with desalination units or transportation systems need to be developed. Power electronics have come out as one of the enablers in order to successfully integrate renewable energy technology to the power systems and therefore improve energy harvesting through dedicated controls. A more detailed investigation should be focused on material development, device design as well as stability of the energy harvesting or storage systems in order to sustain the ongoing energy revolution. For example, carbonaceous materials are abundant on earth and boast of a myriad of allotropic systems that can offer morphological advantages and selectivity for a particular energy system; they therefore have the potential to be used in a wide range of energy storage or conversion processes.

REFERENCES

1. A. Hussain, S.M. Arif, M. Aslam, Emerging renewable and sustainable energy technologies: State of the art, *Renew. Sustain. Energy Rev.* 71, 2017, 12–28. doi: 10.1016/j.rser.2016.12.033
2. D. Larcher, J.M. Tarascon, Towards greener and more sustainable batteries for electrical energy storage, *Nat. Chem.* 7, 2015, 19–29. doi: 10.1038/nchem.2085
3. M.A. Eltawil, Z. Zhengming, L. Yuan, A review of renewable energy technologies integrated with desalination systems, *Renew. Sustain. Energy Rev.* 13, 2009, 2245–2262. doi: 10.1016/j.rser.2009.06.011
4. C. Yin, L.A. Rosendahl, S.K. Kær, Grate-firing of biomass for heat and power production, *Prog. Energy Combust. Sci.* 34, 2008, 725–754. doi: 10.1016/j.pecs.2008.05.002
5. H. Zhao, Q. Wu, S. Hu, H. Xu, C.N. Rasmussen, Review of energy storage system for wind power integration support, *Appl. Energy* 137, 2015, 545–553. doi: 10.1016/j.apenergy.2014.04.103
6. V. Siva Reddy, S.C. Kaushik, K.R. Ranjan, S.K. Tyagi, State-of-the-art of solar thermal power plants – A review, *Renew. Sustain. Energy Rev.* 27, 2013, 258–273. doi: 10.1016/j.rser.2013.06.037
7. REN21, *Renewable Energy Policy Network for the 21st Century. Renewables 2014.* Global Status Report, 2016.

8. United Nation Development Programme (UNDP), *World Energy Assessment 2000 – Energy and the Challenge of Sustainability*. UNDP, New York, NY, 2000 (ISBN 9211261260).
9. M. Carmo, D.L. Fritz, J. Mergel, D. Stolten, A comprehensive review on PEM water electrolysis, *Int. J. Hydrogen Energy* 38, 2013, 4901–4934. doi: 10.1016/j.ijhydene.2013.01.151
10. General Electric PV Systems, 2011, available at: http://www.ge.com/productsservices/energy.html
11. European Wind Energy Association (EWEA), *Deep Water – The Next Step for Offshore Wind Energy*, July 2013.
12. H. Ibrahim, A. Ilinca, J. Perron, Energy storage systems – Characteristics and comparisons, *Renew. Sustain. Energy Rev.* 12, 2008, 1221–1250. doi: 10.1016/j.rser.2007.01.023
13. X. Luo, J. Wang, M. Dooner, J. Clarke, Overview of current development in electrical energy storage technologies and the application potential in power system operation, *Appl. Energy* 137, 2015, 511–536. doi: 10.1016/j.apenergy.2014.09.081
14. A. Poullikkas, A comparative overview of large-scale battery systems for electricity storage, *Renew. Sustain. Energy Rev.* 27, 2013, 778–788. doi: 10.1016/j.rser.2013.07.017
15. P. Simon, Y. Gogotsi, B. Dunn, Where do batteries end and supercapacitors begin?, *Science* 343, 2014, 1210–1211. doi: 10.1126/science.1249625
16. P. Simon, Y. Gogotsi, Materials for electrochemical capacitors, *Nat. Mater.* 7, 2008, 845–854. doi: 10.1038/nmat2297
17. M. Aneke, M. Wang, Energy storage technologies and real life applications – A state of the art review, *Appl. Energy* 179, 2016, 350–377. doi: 10.1016/j.apenergy.2016.06.097
18. Top 10 Lithium-ion Battery Manufacturers in the World, *ELE Times*, 2018. Available at: https://www.eletimes.com/top-10-lithium-ion-battery-manufacturers-in-the-world
19. Y. He, W. Chen, C. Gao, J. Zhou, X. Li, E. Xie, An overview of carbon materials for flexible electrochemical capacitors, *Nanoscale* 5, 2013, 8799. doi: 10.1039/c3nr02157b
20. W. Raza, F. Ali, N. Raza, Y. Luo, K.H. Kim, J. Yang, S. Kumar, A. Mehmood, E.E. Kwon, Recent advancements in supercapacitor technology, *Nano Energy* 52, 2018, 441–473. doi: 10.1016/j.nanoen.2018.08.013
21. C. Zhu, J. Zhai, D. Wen, S. Dong, Graphene oxide/polypyrrole nanocomposites: One-step electrochemical doping, coating and synergistic effect for energy storage, *J. Mater. Chem.* 22, 2012, 6300–6306. doi: 10.1039/c2jm16699b
22. S. Bose, T. Kuila, M.E. Uddin, N.H. Kim, A.K.T. Lau, J.H. Lee, In-situ synthesis and characterization of electrically conductive polypyrrole/graphene nanocomposites, *Polymer (Guildf)* 51, 2010, 5921–5928. doi: 10.1016/j.polymer.2010.10.014
23. D. Qi, Y. Liu, Z. Liu, L. Zhang, X. Chen, Design of architectures and materials in In-plane micro-supercapacitors: Current status and future challenges, *Adv. Mater.* 29, 2017, 1–19. doi: 10.1002/adma.201602802
24. T. Purkait, G. Singh, N. Kamboj, M. Das, R.S. Dey, All-porous heterostructure of reduced graphene oxide–polypyrrole – Nanoporous gold for a planar flexible supercapacitor showing outstanding volumetric capacitance and energy density, *J. Mater. Chem. A.* 6, 2018, 22858–22869. doi: 10.1039/C8TA07627H
25. K. Nomura, H. Nishihara, N. Kobayashi, T. Asada, T. Kyotani, 4.4 V supercapacitors based on super-stable mesoporous carbon sheet made of edge-free graphene walls, *Energy Environ. Sci.* 12, 2019, 1542–1549. doi: 10.1039/C8EE03184C
26. K. Leng, F. Zhang, L. Zhang, T. Zhang, Y. Wu, Y. Lu, Y. Huang, Y. Chen, Graphene-based Li-ion hybrid supercapacitors with ultrahigh performance, *Nano Res.* 6, 2013, 581–592. doi: 10.1007/s12274-013-0334-6
27. A. Louwen, W.G.J.H.M. van Sark, A.P.C. Faaij, R.E.I. Schropp, Re-assessment of net energy production and greenhouse gas emissions avoidance after 40 years of photovoltaics development, *Nat. Commun.* 7, 2016, 13728. doi: 10.1038/ncomms13728

28. H. Li, L. Cheng, Y. Xia, A hybrid electrochemical supercapacitor based on a 5 V Li-Ion battery cathode and active carbon, *Electrochem. Solid-State Lett.* 8, 2005, A433. doi: 10.1149/1.1960007
29. M.B. Sassin, C.N. Chervin, D.R. Rolison, J.W. Long, Redox deposition of nanoscale metal oxides on carbon for next-generation electrochemical capacitors, *Acc. Chem. Res.* 46, 2013, 1062–1074. doi: 10.1021/ar2002717
30. G. Zhu, P. Bai, J. Chen, Z. Lin Wang, Power-generating shoe insole based on triboelectric nanogenerators for self-powered consumer electronics, *Nano Energy.* 2, 2013, 688–692. doi: 10.1016/j.nanoen.2013.08.002
31. Z. Li, J. Chen, H. Guo, X. Fan, Z. Wen, M.-H. Yeh, C. Yu, X. Cao, Z.L. Wang, Triboelectrification-enabled self-powered detection and removal of heavy metal ions in wastewater, *Adv. Mater.* 28, 2016, 2983–2991. doi: 10.1002/adma.201504356
32. R. Elbersen, W. Vijselaar, R.M. Tiggelaar, H. Gardeniers, J. Huskens, Fabrication and doping methods for silicon nano- and micropillar arrays for solar-cell applications: A review, *Adv. Mater.* 27, 2015, 6781–6796. doi: 10.1002/adma.201502632
33. W. Zeng, X.M. Tao, S. Chen, S. Shang, H.L.W. Chan, S.H. Choy, Highly durable all-fiber nanogenerator for mechanical energy harvesting, *Energy Environ. Sci.* 6, 2013, 2631–2638. doi: 10.1039/c3ee41063c
34. D.M. Chapin, C.S. Fuller, G.L. Pearson, A new silicon p-n junction photocell for converting solar radiation into electrical power, *J. Appl. Phys.* 25, 1954, 676–677. doi: 10.1063/1.1721711
35. Global Market Outlook for Solar Power, Africa-EU Renewable Energy Cooperation Programme (RECP). 3-78, 2018.
36. S. Almosni, A. Delamarre, Z. Jehl, D. Suchet, L. Cojocaru, M. Giteau, B. Behaghel et al., Material challenges for solar cells in the twenty-first century: Directions in emerging technologies, *Sci. Technol. Adv. Mater.* 19, 2018, 336–369. doi: 10.1080/14686996.2018.1433439
37. InterSolarEurope, Global Market Outlook – InterSolarEurope. 81, 2018.
38. H. Price, E. Lüpfert, D. Kearney, E. Zarza, G. Cohen, R. Gee, R. Mahoney, Advances in parabolic trough solar power technology, *J. Sol. Energy Eng.* 124, 2002, 109. doi: 10.1115/1.1467922
39. F. Blaabjerg, D.M. Ionel, Renewable energy devices and systems-state-of-the-art technology, research and development, challenges and future trends, *Electr. Power Components Syst.* 43, 2015, 1319–1328. doi: 10.1080/15325008.2015.1062819
40. S. Sundararagavan, E. Baker, Evaluating energy storage technologies for wind power integration, *Sol. Energy* 86, 2012, 2707–2717. doi: 10.1016/j.solener.2012.06.013
41. International Energy Agency (IEA), *World Energy Outlook 2011*, IEA, Paris, France, 2011.
42. P. Wang, Z. Gao, L. Bertling, Operational adequacy studies of power systems with wind farms and energy storages, *IEEE Trans. Power Syst.* 27, 2012, 2377–2384.
43. Eize de Vries, 'Close up – Vestas V164', 2014, available at: http://www.windpowermonthly.com/article/1211056/close-vestas-v164-80-nacelle-hub
44. M. Biswal, A. Banerjee, M. Deo, S. Ogale, From dead leaves to high energy density supercapacitors, *Energy Environ. Sci.* 6, 2013, 1249–1259. doi: 10.1039/c3ee22325f
45. R. Wang, P. Wang, X. Yan, J. Lang, C. Peng, Q. Xue, Promising porous carbon derived from celtuce leaves with outstanding supercapacitance and CO_2 capture performance, *ACS Appl. Mater. Interfaces* 4, 2012, 5800–5806. doi: 10.1021/am302077c
46. L. Sun, C. Tian, M. Li, X. Meng, L. Wang, R. Wang, J. Yin, H. Fu, From coconut shell to porous graphene-like nanosheets for high-power supercapacitors, *J. Mater. Chem. A.* 1, 2013, 6462–6470. doi: 10.1039/c3ta10897j
47. C. Ruan, K. Ai, L. Lu, Biomass-derived carbon materials for high-performance supercapacitor electrodes, *RSC Adv.* 4, 2014, 30887–30895. doi: 10.1039/c4ra04470c
48. K. Qian, A. Kumar, H. Zhang, D. Bellmer, R. Huhnke, Recent advances in utilization of biochar, *Renew. Sustain. Energy Rev.* 42, 2015, 1055–1064. doi: 10.1016/j.rser.2014.10.074

49. R.K. Gupta, M. Dubey, P. Kharel, Z. Gu, Q.H. Fan, Biochar activated by oxygen plasma for supercapacitors, *J. Power Sources.* 274, 2015, 1300–1305. doi: 10.1016/j.jpowsour.2014.10.169
50. M. Genovese, J. Jiang, K. Lian, N. Holm, High capacitive performance of exfoliated biochar nanosheets from biomass waste corn cob, *J. Mater. Chem. A.* 3, 2015, 2903–2913. doi: 10.1039/C4TA06110A
51. A. Bello, N. Manyala, F. Barzegar, A.A. Khaleed, D.Y. Momodu, J.K. Dangbegnon, Renewable pine cone biomass derived carbon materials for supercapacitor application, *RSC Adv.* 6, 2016, 1800–1809. doi: 10.1039/C5RA21708C
52. H. Zhu, X. Wang, F. Yang, X. Yang, Promising carbons for supercapacitors derived from fungi, *Adv. Mater.* 23, 2011, 2745–2748. doi: 10.1002/adma.201100901
53. Q. Li, D. Liu, Z. Jia, L. Csetenyi, G.M. Gadd, Fungal biomineralization of manganese as a novel source of electrochemical materials, *Curr. Biol.* 26, 2016, 950–955. doi: 10.1016/j.cub.2016.01.068
54. Z. Li, L. Zhang, B.S. Amirkhiz, X. Tan, Z. Xu, H. Wang, B.C. Olsen, C.M.B. Holt, D. Mitlin, Carbonized chicken eggshell membranes with 3D architectures as high-performance electrode materials for supercapacitors, *Adv. Energy Mater.* 2, 2012, 431–437. doi: 10.1002/aenm.201100548
55. W. Qian, F. Sun, Y. Xu, L. Qiu, C. Liu, S. Wang, F. Yan, Human hair-derived carbon flakes for electrochemical supercapacitors, *Energy Environ. Sci.* 7, 2013, 379–386. doi: 10.1039/C3EE43111H
56. R. Satish, A. Vanchiappan, C.L. Wong, K.W. Ng, M. Srinivasan, Macroporous carbon from human hair: A journey towards the fabrication of high energy Li-ion capacitors, *Electrochim. Acta.* 182, 2015, 474–481. doi: 10.1016/j.electacta.2015.09.127
57. T. Purkait, G. Singh, M. Singh, D. Kumar, R.S. Dey, Large area few-layer graphene with scalable preparation from waste biomass for high-performance supercapacitor, *Sci. Rep.* 7, 2017, 1–14. doi: 10.1038/s41598-017-15463-w
58. M. V. Jacob, R.S. Rawat, B. Ouyang, K. Bazaka, D.S. Kumar, D. Taguchi, M. Iwamoto, R. Neupane, O.K. Varghese, Catalyst-free plasma enhanced growth of graphene from sustainable sources, *Nano Lett.* 15, 2015, 5702–5708. doi: 10.1021/acs.nanolett.5b01363
59. E. Raymundo-Piñero, M. Cadek, F. Béguin, Tuning carbon materials for supercapacitors by direct pyrolysis of seaweeds, *Adv. Funct. Mater.* 19, 2009, 1032–1039. doi: 10.1002/adfm.200801057
60. W. N. R. Isahak, M. W. Hisham, M. A. Yarmo, Highly porous carbon materials from biomass by chemical and carbonization method: A comparison study, *J. Chem.* 2013, 2013, 1–6.

12 Future Perspective

A growing population and the aim for energy equity will almost unavoidably result in continued increase in global total energy demand. By 2050, the scenarios can be described depending on the availability of fossil fuels and the impact of climate change caused by the human lifestyle and greenhouse gases. Scientific efforts put forth towards the establishment of all useful and commonly known renewable energy resources such as energy storage (battery and supercapacitor), solar cell, fuel cell and hybrid energy technologies, are described in this book. The challenges, opportunities and prospects for carbon nanomaterials in these clean and sustainable energy systems were clearly analyzed in the previous chapters.

Various new and existing improved directions for future and renewable research are identified. In order to bring about a sustainable shift towards renewable energy, one has to take care of the existing and long-standing challenges and approach favourable future research directions with laws and policies in this field, including the following:

1. Electricity generated from renewable resources must be cost-effective and ensure the cost continues to decrease.
2. The search must continue for new forms of renewable energy for energy harvesting as well as energy storage systems.
3. Besides solar and wind, more emphasis should be given to mechanical energy harvesting to power small electronics.
4. New and/or improved nano-carbon materials and carbon-based hybrid materials with boosted properties should be developed.
5. Improvement and emphasis are needed for harvesting renewable energy from waste materials in a proper mechanistic way.
6. Improved and cost-effective methods must be developed for the synthesis of new materials.
7. Laboratory testing techniques should be advanced to facilitate the research conducted, with more realistic tests of energy devices, test methods and the impact of multi-degrees of freedom, and high-cycle loading regimes.
8. The gap between two existing efficient technologies must be reviewed and hybridised to have proper integration with the aim for less loss of energy.
9. Laboratories should be equipped with the next-generation measurement strategies for testing and improving the performance of devices.
10. More collaboration is needed between academia and industry for the proof-of-concept to product-level development.
11. Robust renewable energy policies should be designed and reviewed.
12. High-risk proposals in the field of clean and renewable energy should be encouraged with the hope to develop an 'energy miracle'.

13. Globally, all government bodies should fix a 10-year resolution and future plan in order to have high societal impact and see results.
14. Periodically renew the existing policies, capital investment, implementation cost and long- and short-term effects of new and renewable energy systems and technologies.
15. Identify regulatory changes in the laws and policies of government and implement with the aim to develop 100% renewable energy for the individual countries.
16. Every human should get renewable, affordably priced electricity.

Index

A

AA stacking, 10
AB stacking, 10
ABC stacking, 10
Acetylene black, 40
Acidification, 1
Activated porous carbon, 161
Active material, 39, 72, 143, 148, 152, 158
AFM (atomic force microscopy), 26, 121
Agricultural residue, 4
Air electrode, 42, 43
Al-air cell, 42
Alkali metal, 40
Allotrope, 7, 67, 69
Allotrope of carbon, 90
Amorphous carbon, 7, 90
Anisotropic, 14, 128, 137
Anisotropic thermal expansion, 128
Anode aluminium foil, 50
Anode material, 40, 144, 145, 153
Anode reaction, 77, 82
Aqueous electrolyte, 57
Arc deposition, 20
Arc discharge, 16, 20, 22, 24, 25
Arc-evaporated material, 24
Armchair track, 10
Asymmetric Supercapacitor, 52, 61
Atmospheric carbon dioxide, 1
Sp^2 atom, 27
Atomic force microscopy, 121
Atomic percentage, 27
Avogadro constant, 99

B

Ballistic transport, 19
G band, 27
D band, 27
Bandgap of a photocatalyst, 103
Bandgap of silicon, 68
Battery-supercapacitor, 141
Li-S battery, 37, 41, 42
Bifunctional electrocatalyst, 43
Bioenergy, 4
Biomass, 2, 85, 157, 161, 162
    based energy, 3
    derived nanocarbon, 55, 144–145, 164
Bioreactor, 5
Biowaste materials, 56, 161
π-bond, 9

Bond distance, 13
Bottom-up approach, 10, 69, 127
Breathing mode, 27
Brillouin zone, 10
Brodie method, 22
Bucky ball (C_{60}), 12
Butler-Volmer equation, 98

C

Calcination process, 85
Cantilever, 26, 135
Carbon
    allotrope, 8
    based nanotechnology, 7
    black, 13, 14, 39, 72, 83, 84
    family, 7
    nanobelt, 14
    nanocone, 14
    nanofiber, 13, 43, 58, 69, 110
    nano-onion, 14, 56, 58
    nanosheets, 40
    nanosphere, 40
    segregation, 21
Carbon black, 13, 14, 17, 19, 83, 84
Carbon nanotube double walled (DWCNTs), 19
Carbon nanotube multiwalled (MWCNTs), 11, 19, 24, 55
Carbon nanotube single walled (SWCNTs), 11, 19, 24, 43, 55
Carbonaceous soot, 25
Carbon/Si heterojunction solar cell, 69
Carrier mobility, 54, 69
Catalytic activity, 26, 82, 85, 90, 91, 97, 100, 109, 112–113
Catalytic property, 41
Catalytic thermal decomposition, 21
Cathode aluminium foil, 50
Cathode material, 39, 144, 153
Cathode reaction, 36, 77
Cation exchange, 59
CCGT (combined cycle gas turbine), 3
CDCs (Carbide-derived carbon), 55
Cell potential, 35
Cellulose, 5, 85, 87
Characterisation, 25, 83
Charcoal, 13
Charge carrier, 104, 113, 123
Charge transport carrier, 73
Chemical doping, 75, 163
Chemical energy, 35, 77, 91, 119, 148

169

Chemical inertness, 19
Chemical potential, 35
Chemical reduction, 11, 84
Chemical synthesis, 22
CVD (Chemical vapor deposition), 20, 25, 56, 69, 84, 101, 124, 133, 137
Chiral
  indice, 11
  vector, 11
Chronoamperometric, 99
Chronopotentiometric, 99
CO_2 adsorption, 110, 113
Coal, 2, 3, 119
CO_2 electroreduction, 109
CO_2 emission, 2
Concentric layered structure, 14
Conducting gel electrolyte, 52
Conducting polymer, 51, 55, 58
sp^2 conjugated carbon atoms, 37
Conjugated molecule, 13
Contact mode, 26
Contact-mode AFM, 123
Conventional energy source, 114
Conversion chemistry, 37
Corrosion problem, 61
Corrosive salt, 55
Coulombic efficiency, 40, 42
Crystalline, 7, 69, 74, 115, 128, 137, 153, 159
Crystalline silicon, 67, 69, 74
Current density, 19, 80, 97, 98, 99, 104, 110, 111
Cyclability, 42
Cyclic voltammetry, 53, 98, 100, 110
Cycling stability, 55, 82

D

Decoupled, 19
Defect, 27, 69, 73, 90, 91, 104, 110
Deforestation, 1
Desorption, 22, 95, 108, 110
Diamond, 7, 15, 90, 110
Dichalcogenide, 39
Diffuse layer, 50
Diffusion coefficient, 80, 86
Dirac equation, 10
Dirac points, 10
Disordered carbon nanotube, 42
Dissociation, 21
Divacancy defects, 41
N-doped graphene, 22, 24, 84, 110, 115, 116
  sulfur (NGS) composites, 42
Doped graphitic sample, 27
Dry separator, 49
Dye-sensitized solar cell, 147, 160

E

Earth-abundant cathode, 37
Effective radius, 35

Electrical energy, 35, 120, 128, 141, 145, 148, 149
Electric double-layer effect, 131
Electricity production, 2, 159
Electrocatalytic applications, 85
Electrocatalyst and photocatalyst, 107, 114
Electrochemical and photochemical, 106
Electrochemical as well as thermal routes, 157
Electrochemical behavior of pt nanoparticles, 86
Electrochemical bio-capacitors, 148, 155
Electrochemical catalyst, 87
Electrochemical cells, 150
Electrochemical characterization, 83
Electrochemical characteristic, 143
Electrochemical charge storage, 153
Electrochemical CO_2 reduction, 112, 117
Electrochemical degradation of organic pollutant, 138
Electrochemical double-layer capacitor, 158
Electrochemical doping, 163
Electrochemical energy storage, 17, 49, 62, 158, 159, 161
Electrochemical flow capacitor, 144
Eletrochemically
  active manganese oxide, 145
  active surface area, 83
  splitting water, 101
  treated to avoid, 101
Electrochemical magnesium deposition\
  dissolution, 36
Electrochemical material, 165
Electrochemical methanol oxidation, 83
Electrochemical or photoelectrochemical device, 150, 161
Electrochemical oxidation and reduction reaction, 145
Electrochemical potential, 35, 130
Electrochemical process, 24, 133
Electrochemical properties of negative electrodes, 152
Electrochemical reaction, 49, 97, 98, 142
Electrochemical reduction, 11, 23, 31, 57, 78, 108, 116, 117
Electrochemical storage device, 141
Electrochemical supercapacitor, 61–64, 147, 152, 164, 165
Electrochemical systems, 113, 147, 150
Electrochemical water splitting, 93, 115, 155, 161
Electrode-electrolyte interface, 108
Electrolytic capacitor, 49
Electromagnetic effect, 121
Electromagnetic energy harvesters, 121
Electron beam, 26
Electron delocalisation, 71
Electron diffraction pattern, 26
Electronegative absorption, 14
Electron hole creation mechanism, 68
Electron-hole pair, 110, 113
Electronic charge, 35, 77

Index

Electronic conductor, 35
Electronic state, 104
Electronic structure, 10, 84, 100
Electronic transport, 35
Electrophoretic deposition, 57
Electropositive material, 37
Electrospun, 14, 110, 128
Electrospun polyacrylonitrile, 110
Electrostatic capacitor, 49
Electrostatic effect, 121
Electrostatic micro power generator, 122
Energy barrier, 67
Energy conversion, 148, 161
   and generation, 149
   and piezotronic, 136
   and storage, 7, 17, 44, 49, 63, 68, 75, 87, 115, 147, 151
   device, 77, 122, 161
   efficiency, 124, 131, 135, 136
   efficiency and robustness, 131
Energy conversion efficiency, 124, 131, 135, 136
   and robustness, 131
   of the TENG, 131
Energy efficiency, 6, 50, 109
Energy harvesting device, 123
Energy intensity, 2
Energy storage technology, 158, 163, 164
Environmental Pollution, 3, 89, 119
Epitaxial growth, 20, 22
Euler's polyhedron, 12
European Union population, 157
Excitation, 21, 142
Exchange current density, 97, 98, 99

F

Fast charging, 41, 49, 60, 159
Faraday's constant, 35, 98
Faraday's law of induction, 121
Faradic efficiency, 93, 97, 99, 109–110, 113
Faradic loss, 99
Flexible self-charging nanogenerator, 146
Fossil energy, 3
FTIR (Fourier transform infrared), 26

G

GCD (Galvanostatic charge/discharge), 53
Gas constant, 98
Gas diffusion layer, 77
Gaseous biofuels, 4
Geothermal energy, 3
Global
   capacity, 4, 161
   carbon footprint, 162
   co_2 emission, 157
   coverage, 159

energy, 1, 3, 167
energy economy, 3
installation, 159
market, 68, 160, 164
problem, 141
status, 162
warming, 1, 89, 106, 114, 119
Gouy and Chapman, 50
Graphene
   composite, 10, 11, 69, 145
   flake, 20, 22, 73
   sulfur nanocomposite, 42
   waste derived, 23
Graphene/graphene oxide heterojunction, 104
Graphene/Si heterojunction solar cell, 69
Graphene-wrapped sulfur particle, 42
(GO) Graphite oxide, 22, 26
Graphitisation, 14
Gravimetric capacitance, 59
Green-house effect, 1
Greenhouse gas emission, 1, 77, 161, 163
Grid, 3, 35
   integration, 4
   system, 157, 162

H

Harvesting biomechanical energy, 146
Helicity, 15
Helmholtz, 50
Hemicellulose, 85
Heteroatom doped graphene, 69
Heterogeneous catalysis, 19
Hexagonal lattice, 26
2D hexagonal sheet, 10
Heyrovsky reaction, 95
High electron transport, 67
High-energy storage system, 37, 60
Highly hydrophilic, 22
HOPG (Highly oriented pyrolytic graphite), 19, 20, 23
High Power Density, 49, 52, 60, 80, 144
High-power lithium battery, 39
High-strength engineering fiber, 19
Hofmann method, 22
Hole transfer layer, 67
Honeycomb, 9, 10, 19
Hopping, 10
Human-machine interfacing, 132
Hummers method, 22
HEVs (hybrid electric vehicles), 39, 141, 158
Hybrid energy technology, 151, 167
sp^2-hybridized carbon, 16, 84, 90
Hybrid Na-ion capacitor, 145
Hybrid nanogenerator/biofuel cell, 151
Hybrid sodium ion capacitor, 23
Hydrogen bonding, 10
Hydrogen evolution rate, 104

Hydrogen generation, 91, 112
Hydrogen plasma treatment, 21
Hydrogen storage, 14
Hydro power, 2
Hydroquinone, 23
Hydrothermal reduction, 11

I

Inorganic photocatalyst, 114
Insoluble sulphide, 42
π–π interaction, 10
Interface interaction, 43
Interlayer spacing, 27
Intermittency, 4
Intermittent energy density, 49
Intermittent generator, 49
Internal resistance, 52
Ionic charge, 77
Ionic liquid, 43, 61, 159
Ionic transport, 35
Ionisation, 21
Irreversible intercalation, 60

J

Joule-heating, 16

K

Kinematic viscosity, 80
Koutecky-Levich equation, 80
K-point, 10

L

Laser ablation, 25
Lattice vector, 11
Lignin, 85, 87
Linear sweep voltammetry, 97, 100
Liquid biofuels, 4
Lithium, 141
    diffusion, 39
    ion batteries, 9, 11, 35, 36, 37, 44, 45, 61, 141, 158
    ion capacitor, 144, 152
    ion deintercalation, 39
    ion hybrid supercapacitor, 159
    ion–intercalated graphite, 35, 37
    ion intercalation, 39
    ion technology, 41
    oxygen battery, 148
    solid-state diffusion, 39
Lithium-ion hybrid supercapacitor, 159
Lithography, 53, 69, 133, 137
Low cost material, 59
Low Energy Density, 61, 141, 159

M

Magnesium anode, 42
Magnesium-ion (Mg-ion) batteries, 36, 41, 44
Mass/charge transport, 43
Maximum symmetric molecule, 13
Maxwell's equations, 126
Mechanical exfoliation, 19, 20, 24, 56
Mechanical strength, 57, 74, 82, 142
Mechanical vibration energy, 119
Mesoporous graphene foam, 100
Metal-air batteries, 37, 42, 44, 149, 150
Metal-based catalyst, 78, 82, 85, 90
    Cathode, 82
    Electrocatalyst, 161
    ORR catalyst, 85
Metal-free
    bifunctional catalysts, 47, 86
    carbonaceous electrocatalyst, 114
    catalyst, 26, 78
    current collector, 122
    efficient photocatalyst, 116
    electrocatalyst, 85
    electrochemical catalysts, 87
    graphene, 43, 47
    hydrogen evolution catalyst, 115
    nanocarbon-based electrocatalyst, 84
    oxygen reduction, 87
    polymeric photocatalyst, 117
Metal ion, 35, 164
Metal organic frameworks, 90
Metal-sulfur battery, 37, 41, 42, 44
Mg-air cells, 42
Mg-carbon composite, 42
Micro-batteries, 53
Micro-electromechanical systems, 119
Microorganism, 5, 85, 87
Microporosity, 13
Mobility, 19, 69, 90
Molecular wires, 19
Monocrystalline, 20
Monovalent ions, 35

N

Nanocarbon material, 40, 44, 78, 82, 85
Nanodiamonds, 58
Nano-energy cell, 146, 154
Nanofibrous membrane, 134
Nano-form carbon, 7
Nanohybrid capacitor, 144, 153
Natural gas, 2, 3, 119
N-doped
    and P-doped graphene, 105
    carbon, 85
    carbon nano fiber aerogel, 87, 110
    graphene, 22, 24, 29, 31, 32, 42, 47, 84, 110, 115

Index

nanodiamond, 110
reduced graphene oxide, 85
Neutral polymer, 58
Nickel substrate, 21
Non faradic process, 51, 52
Nuclear power, 2

O

Open-circuit voltage, 22, 124, 133
Optical transmittance, 19
Optical transparency, 57
Organic light-emitting diodes, 71
Organic liquid electrolyte, 41, 56
Organic solar cell, 72, 75
Overtone, 27
OER (oxygen evolution reaction), 43, 92, 94, 96–100
ORR (oxygen reduction reaction), 43, 79–85, 99

P

Peanut shells, 23, 56, 162
PEC water splitting, 105
PECVD (Plasma-enhanced CVD), 21
Perovskite-based solar cell, 73
Phase transformation of the material, 143
E_{2g} phonon, 27
Photocatalytic CO_2 reduction, 106, 117
Photocatalytic reduction of CO_2, 113
Photocatalytic water splitting, 93, 101, 115, 116
Photoelectrocatalysis, 89, 91, 93, 106, 114
Photosynthesis, 92, 101, 105
Photovoltaic, 3–5, 68, 106, 150, 157, 163
  application, 69
  are categorized, 68
  cell, 74, 75
  charge generation mechanism, 142
  device, 67–69
  effect, 67, 74, 142
  principle and method, 74
  property, 69
  technology, 68
  was introduced, 69
Physisorption, 60
Phytic acid, 43
Piezoelectric co-efficients, 122
Piezoelectric effect, 121, 123, 135
Piezoelectric transduction mechanism, 122
Planner supercapacitors, 52
Platinum nanoparticle, 84
PANI (Polyaniline), 51, 58
Polyaniline aerogels, 43
Polyaromatic hydrocarbons, 10
Polycrystalline, 21, 67, 159, 160
PDMS (Polydimethylsiloxane), 22, 128, 131, 132, 133

PET (Poly(ethylene terephthalate)), 22
PMMA (Polymethyl methacrylate), 21
PPy (polypyrrole), 51, 58, 147, 163
PVDF (poly(vinylidene fluoride)), 121, 122, 124, 129, 133
κ-point phonons, 27
Pore accessibility, 56
Pore size distribution, 56
Porous carbon electrode, 50
Porous hollow carbon@sulfur, 42
Porous structure, 14, 41, 55, 90, 100
Porphyrin, 24
Power delivery systems, 60
Power generation technologies, 3
PEM (Proton exchange membrane), 77, 93, 113
  fuel cell, 78, 80, 81, 151
Pseudocapacitance, 51, 60, 143, 159
PTh (polythiophene), 51, 58
Pyrene, 24
Pyroelectric co-efficient, 129
Pyroelectric coupling, 128
Pyrolysis, 127, 137, 165

Q

Quantum confinement, 69, 104
Quantum Hall effect, 10

R

Ragone plot, 54
Raman spectra, 26
Rate capability, 44, 141
Rechargeable battery, 37, 41
Recombination kinetics, 72
Redox active functionalities, 51
Redox potential, 112
Redox reaction, 37, 59
rGO (Reduced graphite oxide), 22, 28, 73, 84, 145, 148
  film, 27, 146
  n-doped, 41
  s-doped, 41
Reduction of CO_2, 106, 108, 110, 112–113, 116–117
Reduction of H_2O, 113
Reduction potential, 106, 107
Renewable energy, 1, 5, 6, 157, 164, 167, 168
  for energy harvesting, 167
  from waste material, 167
  policies, 162, 167
  resources, 3, 91, 92, 93, 106, 167
  system and technology, 168
  technology, 158, 162
Resistivity, 19
Reversible hydrogen electrode, 89
Reversible capacity, 40
Rhombohedral, 10

Rotating disk electrode, 72, 80
Rotational symmetry, 13

S

Saline systems, 42
Sandwich-type or traditional SCs, 52
SEM (scanning electron microscopy), 26, 129
Schottky barrier, 123, 127, 137
SAED (Selected area diffraction pattern), 26
Self-discharge, 39, 61
Self-powered devices, 142, 152
Self-powered energy systems, 119
Sensitiser, 72
Separator, 52, 93, 143, 145
Single-crystal silicon carbide, 22
Single-layer graphene, 10
SWCNHs (Single-wall carbon nanohorns), 83
Small-scale mechanical energy, 159
NIBs (Sodium-ion batteries), 35, 40, 44
Solar cell technology, 67, 68
Solar-to-hydrogen conversion, 90
Solar-to-fuel conversion, 105
Solid carbon, 7
Solid-state diffusion, 40
Solid-state Supercapacitor, 52
Solution-based chemistry, 10
Solution-based self-assembly, 58
Soluble polysulfides, 42
Solvothermal reduction, 11
Specific capacitance, 53, 149
Specific capacity, 40
Specific power, 53
Sporadic mechanical source, 120
Stable aqueous colloids, 22
Standard electrode potential, 40
Standard reduction potential, 110
Staudenmaier method, 22
Stern and Geary, 50
Structural flexibility, 57
Sublimation, 21
Sulfur cathode, 42
Sulfur composite cathode, 41
Super hydrophobicity, 19
Surface charge potential, 132
Surface chemistry, 26
Surface-dependent properties, 7
Surface morphology, 26
Sustainable composites, 37
Sustainable energy, 5, 91, 93, 150, 158, 162, 167
A_1g symmetry, 27
Synergistic effect, 58, 104, 163

T

Tafel plot, 97, 98, 99
Tafel reaction, 95

Tapping mode, 26
TEM (Transmission electron microscopy), 26
Tetraglyme, 42
Tetrahedral, 7
Thermal conductivity, 9, 19
Thermal CVD, 21, 133
Thermal energy, 6, 119–120, 128, 145, 154
Thermal release tape, 21
Thermodynamic equilibrium, 40
Thermodynamic potential, 105
Thermodynamic reduction potential, 107
Thermoelectric, 41, 137, 142, 150, 155
Three-electrode system, 57
Tip-surface interactions, 26
Top-down approach, 10, 69, 70, 127
Transition metal oxide, 51
Triboelectric charge density, 131
Trigonal planar, 7
Turnover frequency, 99

U

Ultrasonication, 20
Ultrasonic waves, 124, 132, 135
UV-vis (Ultraviolet-visible), 26
Undoped graphitic sample, 27
Unit cell, 10

V

Vibrational energy harvesting, 135
Vibration-to-electricity conversion, 121
Volmer reaction, 94–96
Volumetric capacity, 41, 59
Volumetric energy densitiy, 53, 159

W

Wind
 energy, 3, 4, 152, 160, 161
 turbines, 4

X

XRD (X-ray diffraction), 26, 27, 28
XPS (X-ray photoelectron spectroscopy), 26, 27, 28

Z

Zero-strain, 41
Zigzag tracks, 10
Zn-air cell, 42, 43
Zone boundary, 10